Laboratory Guide for

An Introduction to Physical Science

Eighth Edition

James T. Shipman

Ohio University

Clyde D. Baker

Ohio University

Houghton Mifflin Company   Boston   New York

Editor-in-Chief: Kathi Prancan
Assistant Editor: Marianne Stepanian
Editorial Assistant: Katie Shutzer
Senior Manufacturing Coordinator: Lisa Merrill
Executive Marketing Manager: Karen Natale

Cover Design: Harold Burch, Harold Burch Designs, New York City
Cover Photographs: National Aeronautics and Space Administration

Printed in the U.S.A.

ISBN: 0-669-41718-1

1 2 3 4 5 6 7 8 9–PO–00 99 98 97 96

Experiment 40   Locating Stars in the Night Sky   *285*
Experiment 41   Motions and Phases of the Moon   *299*
Experiment 42   Observing the Phases of the Moon   *305*
Experiment 43   Hubble's Law   *313*
Experiment 44   Air Pressure   *321*
Experiment 45   Humidity   *327*
Experiment 46   Weather Maps (Part 1)   *333*
Experiment 47   Weather Maps (Part 2)   *343*
Experiment 48   Topographic Maps   *349*
Experiment 49   Minerals   *355*
Experiment 50   Rocks   *367*
Experiment 51   Rock-Forming Minerals   *375*
Experiment 52   Igneous Rocks and Crystallization   *379*
Experiment 53   Sedimentary Rocks   *385*

Appendix I   Measurement and Significant Figures   *389*
Appendix II   Conversion of Units   *391*
Appendix III   Experimental Error   *393*
Appendix IV   Conversion Factors   *395*

Contents

Preface   *v*

Laboratory Safety   *vi*

Experiment 1   Graphs   *1*

Experiment 2   Measurement   *11*

Experiment 3   The Simple Pendulum   *21*

Experiment 4   Uniform and Accelerated Motion   *31*

Experiment 5   Determining g, the Acceleration of Gravity   *39*

Experiment 6   Newton's Second Law   *47*

Experiment 7   Hooke's Law for a Vibrating Spring   *53*

Experiment 8   Centripetal Acceleration and Force   *59*

Experiment 9   Laws of Equilibrium   *65*

Experiment 10   Waves   *73*

Experiment 11   Interference of Light Waves   *83*

Experiment 12   Mirrors, Lenses, and Prisms   *91*

Experiment 13   The Refracting Telescope   *103*

Experiment 14   Color   *111*

Experiment 15   Static Electricity   *119*

Experiment 16   Magnetism and Electromagnetism   *127*

Experiment 17   Ohm's Law   *135*

Experiment 18   Electromagnetic Waves   *143*

Experiment 19   Temperature   *153*

Experiment 20   Specific Heat   *157*

Experiment 21   Heat of Fusion   *163*

Experiment 22   Heat of Vaporization of Water   *167*

Experiment 23   Inverse-Square Law of Radiation   *173*

Experiment 24   The Rutherford Scattering Box   *179*

Experiment 25   Radiation   *191*

Experiment 26   Spectroscopy   *197*

Experiment 27   Density of Liquids and Solids   *205*

Experiment 28   Oxygen   *213*

Experiment 29   Percentage of Oxygen in Potassium Chlorate   *219*

Experiment 30   Percentage of Oxygen and Nitrogen in the Air   *223*

Experiment 31   An Exothermic Chemical Reaction   *227*

Experiment 32   Avogadro's Number   *233*

Experiment 33   Molecular Structure   *237*

Experiment 34   Solutions and Solubility   *245*

Experiment 35   Pressure-Volume Relationship of Gases   *251*

Experiment 36   Chemical Qualitative Analysis   *259*

Experiment 37   Chemical Quantitative Analysis (Volumetric)   *263*

Experiment 38   Kepler's Law   *267*

Experiment 39   Stars and Their Apparent Motions   *273*

Preface

Laboratory work, with its practical applications, helps familiarize students with nature's physical laws. Students often have difficulty visualizing the physical concepts discussed in lectures, and practice in the laboratory with apparatus and raw materials clarifies the picture. The experience of solving physical problems by applying knowledge gained in the classroom to laboratory activity is highly rewarding. However, the benefits derived from laboratory work are in direct proportion to students' interest, curiosity, and keenness of observation. The laboratory simply provides the means for carrying out experiments.

Students should approach the laboratory experiments with a desire for understanding and a willingness to learn. They will profit from critically evaluating the results of each experiment with an open mind and a commitment to discovering the truth.

The *Laboratory Guide* contains fifty-three experiments in the five major divisions of physical science: physics, chemistry, astronomy, geology, and meteorology. Each experiment includes an introduction, learning objectives, a list of apparatus, procedures for taking data, and questions. In addition, many experiments call for calculations and the plotting of graphs, and this guide provides space and graph paper for those purposes.

The *Instructor's Resource Manual* now includes data for many experiments, plus calculations using the data provided. These data and calculations will be a great help to the inexperienced laboratory assistant, especially the assistant doing the grading.

Student backgrounds often vary considerably. Many students are familiar with plotting graphs, while others have difficulty reading and interpreting graph data. Experiment 1, Graphs, can be done first or later in the course, or omitted entirely. The other experiments may be completed in any order.

We acknowledge the contributions of many students and colleagues who have contributed to this and previous editions of the *Laboratory Guide*. Our thanks to them and to everyone else who has sent us comments and suggestions. A special thanks to Karen Baker at Ohio University, and to the staff at Houghton Mifflin Company.

J. T. S
C. D. B

Laboratory Safety

Safety in the laboratory is the responsibility of both the instructor *and every student*. The instructor should teach safety at all times, and the student should follow all instructions, both verbal and printed.

The laboratory should have adequate first aid equipment, and personnel with the knowledge to use the equipment, in case of injury. Fire extinguishers should be in view and in good working order, proper ventilation systems installed, safe methods readily available for the disposal of waste chemicals and other materials, and all electrical equipment examined for shock hazards.

The experiments in this laboratory guide are designed to be done safely, but students should exercise caution at all times. Most accidents in the laboratory result from carelessness and failure to take proper precautions in carrying out experiments. *Do not perform any unauthorized experiments.* Be careful in handling glass and sharp tools. Also, be careful to avoid chemical spills and splashing hot liquids that can cause serious eye and skin injuries and damage clothing. Spills on the floor of the laboratory may produce a slippery surface, setting the stage for falls. When handling electrical equipment, be extra careful, in order to avoid electrical shock.

Before touching any laboratory experiment to perform an experiment, wait until the instructor has explained safety precautions and has given permission to start the experiment.

When performing experiments with chemicals, never use more of a chemical than is called for in the experiment. Never pour a liquid from a bottle into a test tube; instead use a piece of glazed paper to transfer the solid to the test tube. As a precaution against contamination, never return unused chemicals to the supply bottle or container.

All electrical experiments using a source of electrical energy must be checked by the instructor before they are connected to the source of energy. This is for the safety both of the student and of the delicate electrical measuring instruments.

If you have any doubts about the safety precautions to be taken when performing an experiment, ask the instructor for assistance.

Graphs

INTRODUCTION

A **graph** is a pictorial representation of ordered pairs of numbers from which the reader may quickly determine relationships between quantities. In many of the experiments performed in this laboratory guide, quantities that change in value will be studied. A change in one quantity may cause another quantity to change. We say one quantity is a function of the other. For example, the area of a circle is a function of the radius. That is, the area depends on the size of the radius. When we increase the size of the radius the area of the circle increases accordingly. In this example the radius is the independent variable and the area is the dependent variable.

After the ordered pairs of numbers [for example (x_1, y_1), (r_1, A_1)] are plotted on graph paper, the plotted points are connected by a smooth line. The line that is drawn may be straight or it may be curved. The general name curve is used in reference to all graphs whether the line is actually curved or straight.

LEARNING OBJECTIVES

After completing this experiment, you should be able to do the following:

1. Define and explain the term graph.
2. Plot a graph of a linear relationship using a mathematical equation or formula to obtain ordered pairs of numbers.
3. Plot a graph of a nonlinear relationship using a mathematical equation or formula to obtain ordered pairs of numbers.
4. Distinguish between dependent and independent variables.
5. Define slope, determine the slope of a straight line, and give a physical interpretation of the slope.
6. State the equation for a straight line that passes through the origin in terms of the plotted variables.
7. State the equation for a parabola.
8. Replot a parabola as a straight line.

APPARATUS

Three sheets of linear graph paper, pen or sharp pencil, ruler. (A hand calculator is useful but not necessary.)

PROCEDURE 1

A graph is a picture of ordered pairs of numbers. Thus, to draw a graph we must use graph paper and a set of ordered pairs of numbers. Graph paper is provided at the end of this experiment. You must calculate or determine experimentally the ordered pairs of numbers. In this experiment, calculate the ordered pairs using the mathematical equation or formula for the circumference of a circle as a function of the diameter of the circle (Eq. 1.1), then record the values in Data Table 1.1. Values for the circumference should be rounded off to two decimal places. The circumference of a circle equals the value pi (π) times the diameter of the circle. This can be written in symbol notation as

$$C = \pi d \tag{1.1}$$

where C = the circumference,

$\quad\quad d$ = the diameter,

$\quad\quad \pi$ = 3.14.

Data Table 1.1

Diameter (d) centimeters (cm) Independent variable	Circumference (C) centimeters (cm) Dependent variable	Ratio C/d
1		
2		
3		
4		
5		
6		
7		
	Average value of C/d	

Plot a graph [Graph number one] of the circumference of a circle as a function of the diameter using the information in Data Table 1.1. The following guidelines should be used when plotting graphs:

1. Make the proper choice of graph paper.
2. Use a pen or a sharp pencil.
3. Write neatly and legibly.
4. Choose scales so that the major portion of the graph paper is used.
5. Choose scales for the x and y axes that are easy to read and plot.
6. Plot the independent variable on the horizontal or x axis and the dependent variable on the vertical or y axis.

The two quantities (circumference and diameter) plotted for this experiment are called *variables*. For the data given, the diameter of the circle is the independent variable and the circumference the dependent variable.

7. Plot each ordered pair of numbers as a single point (d_1, C_1), (d_2, C_2), etc.
8. Plot each point clearly using a dot surrounded by a small circle ⊙.
9a. Label the x and y axes with the quantity plotted. Examples: distance, time, area, volume, etc.
9b. State the units of each quantity plotted. Examples: centimeters, seconds, cm^2, cm^3, etc.
10. Draw a smooth line connecting the plotted points. "Smooth" suggests that the line does not have to pass exactly through each point but connects the general areas of significance.
11. Give the graph a title and place the title on the graph (usually upper center of graph). The name of the graph is taken from the labels of the x and y axes plus reference to the object to which the data refer. Examples: volume versus radius for a sphere; period versus length for a simple pendulum.
12. Place your name and the date on the graph (usually lower right side of graph). See Fig. 1.1 for an example of a graph plotted according to the above instructions.
13. Determine the slope of the curve. The slope of a curve (for this graph the curve is a straight line) is defined as the change in the y or vertical values divided by the change in the x or horizontal values. This can be stated in symbol notation as $\Delta y/\Delta x$. Read this as "delta y over delta x" (delta means "change in"). Using m as the symbol for the slope we can write

$$\text{Slope} = m = \frac{\Delta y}{\Delta x} = \frac{y_2 - y_1}{x_2 - x_1}$$

If the straight line passes through the origin (ordered pair, 0, 0) then we can write

$$Slope = m = \frac{y_2 - y_1}{x_2 - x_1} = \frac{y_2 - 0}{x_2 - 0} = \frac{y_2}{x_2}$$

$$m = \frac{y}{x}$$

or
$$y = mx$$

This is the general equation for a straight line that passes through the origin. Remember, the slope (m) of a straight line has a constant value. Therefore, the values (y_2, x_2) can be any ordered pair of points on the plotted graph.

The point where the curve crosses the y axis is known as the y intercept. In the equation above, when x is zero, y is zero. Thus the curve crosses the y axis at the origin. Later we shall have curves that do not pass through the origin, and the y intercept will have a numerical value.

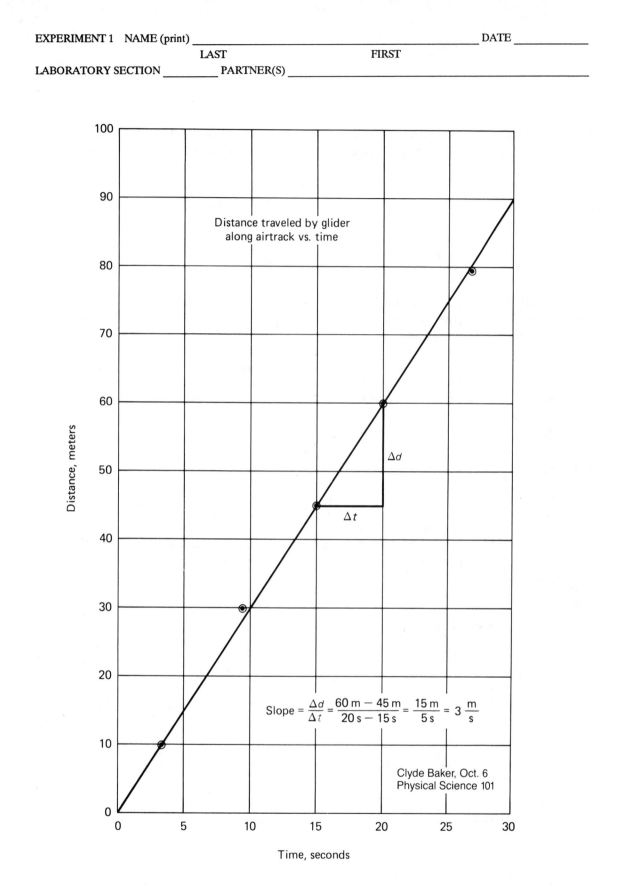

Figure 1.1

PROCEDURE 2

The relationship between the area of a circle and the radius of the circle can be stated as follows: The area of a circle equals the constant pi (π) times the radius squared, or written in symbol notation,

$$\text{Area} = \pi(\text{radius})^2$$

or

$$A = \pi R^2 \tag{1.2}$$

where A = the area in square meters,

R = the radius in meters,

π = 3.14.

Using Eq. 1.2, calculate ordered pairs of numbers for seven values of the radius. Record the calculated values in Data Table 1.2. Plot a graph [Graph number two] of the area (y axis) as a function of the radius (x axis). Label the graph completely.

The equation $y = kx^2$ is the general equation for a curve called a **parabola**. In this equation y is a function of x squared (x^2) where, as in the equation for a straight line ($y = mx$), y is a function of the linear value of x. When the relationship between pressure (p) and volume (V) for an ideal gas (Boyle's law) is studied, you will be asked to plot the equation $pV = k$. This is the equation for a hyperbola. Stated in terms of x and y, the equation is $y = k/x$. See Part 2, Section 8.4 in the Study Guide for more information on the parabola and the hyperbola.

PROCEDURE 3

Many times the data plotted from an experiment in the laboratory yield a curve whose true nature is difficult to observe because of limited data. The true nature (the equation of the curve) can be determined when the data are extended and replotted in order to obtain a straight line. The third column in Data Table 1.2 is labeled (R^2). Calculate the seven values for the radius squared and record them in Data Table 1.2. Plot a graph [Graph number three] with the area as a function of the radius squared. Label the graph completely and calculate the slope.

Data Table 1.2

Radius (R) meters	Area (A) (meters)^2	R^2 (meters)^2
1		
2		
3		
4		
5		
6		
7		

QUESTIONS

1. Describe the method you used to determine the magnitude of the divisions for the *x* and *y* axes in order to plot a graph that uses the major portion of the graph paper.

2. Distinguish between dependent and independent variables.

3. Name the dependent and independent variables in Procedure 2.

4. How is the name for the graph selected?

5. How are the units for the slope determined?

6. What is the relationship between the units for the *x* and *y* axes and the units for the slope?

7. Is the relationship between the circumference and the diameter of a circle a linear relationship? Explain your answer.

8. What is the relationship between the slope of the curve in Procedure 1 and the ratio of C/d?

9. When the value for *x* increases faster than the values for *y*, which way (toward or away from) will the graph curve with respect to the *y* axis?

10. State in words the relationship between the area of a circle and the radius^2.

Measurement

INTRODUCTION

All experiments to be performed in this laboratory require one or more measurements. A **measurement** is defined as the ratio of the magnitude (how much) of any quantity to a standard value. The magnitude is first determined by using some device to compare the unknown quantity to a standard and then expressing the comparison as a number. A number representing the magnitude is not sufficient to express the measurement, however; a unit must also be assigned to the number. To record a person's height as 64 is meaningless. We must record it as 64 inches. All measurements, therefore, must be recorded by number and unit.

Many devices (rulers, balances, clocks, speedometers, thermometers, voltmeters, oscilloscopes, spectrometers, etc.) are used to make measurements. The information obtained from these instruments must be recorded and evaluated in order to obtain a truer value of the properties of the physical quantity. Such terms as **least count, significant figures, precision, accuracy, and percent error** must be learned and used.

Laboratory experiments require the taking and recording of data. Values read from the measuring instrument are to be expressed with numbers known as significant figures. By definition, a **significant figure** is a number that contains all known digits plus one doubtful digit. That is, the significant figure includes all digits as read from the instrument plus one doubtful digit. The doubtful digit is an estimate of the smallest scale division on the instrument. For each measurement in all experiments the student must record significant figures. Once the measurements have been made and recorded, it may be necessary to perform arithmetical operations using the significant figures. In the multiplying or dividing of two or more measurements, the number of significant digits in the final answer can be no greater than the number of significant digits in the measurement with the least number of significant digits. When adding or subtracting, the last digit retained in the sum or difference should correspond to the first doubtful decimal place.[*] Example:

$$
\begin{array}{r}
16.3 \ \text{cm} \\
203.4 \ \text{cm} \\
+ \ \ 4.07 \ \text{cm} \\
\hline
223.8 \ \text{cm}
\end{array}
$$

---

[*] See Appendix I for more information.

The following procedure is usually used to round off significant figures to fewer digits. If the last significant digit on the right is less than 5, drop it and insert zero instead. If the last significant digit is 5 or greater, drop it and increase the preceding digit by one.

Examples

Round off the following numbers to two significant digits.

247   Since the last digit on the right is greater than 5, drop it and increase the preceding digit by one, for the result 250.

243   Since the last digit on the right is less than 5, drop it and insert zero instead, for the result 240.

275   Since the last digit on the right is 5, drop it and increase the preceding digit by one, for the result 280.

A number representing a measured quantity may also be expressed by means of the powers-of-ten notation. To express a number in this way, place the decimal point after the first significant digit, then use a power of ten to locate the true position of the decimal point. For example, the average distance from the Sun to the Earth is 93,000,000 mi. Using powers-of-ten notation, this can be written as 9.3×10^7 mi. Likewise, a very small number can be expressed using the powers-of-ten notation. For example, the thickness of a piece of paper is 0.0001 m. This can be written as 1×10^{-4} m.

Just recording a numerical value for the measured quantity is not sufficient to express a physical quantity. A unit must also be indicated. For example, a measurement is taken for the length of the laboratory table and recorded as 183. This number has no real meaning unless expressed with a unit. A correct recording would be 183 centimeters (cm).

The process of taking any measurement always involves some uncertainty. Such uncertainty is usually called experimental error. Two methods are used to calculate the amount of error. (1) When an accepted or standard value of the physical quantity is known, the percentage error is calculated (in a comparison of the experimental measurement with a standard). (2) When no accepted value exists, a percentage difference is calculated (in a comparison of two or more experimental measurements).

$$\text{Percentage error} = \frac{\text{absolute difference}}{\text{accepted value}} \times 100$$

$$= \frac{(E_v - A_v)}{A_v} \times 100$$

where E_v = experimental value and A_v = accepted value (known as standard value). Absolute difference means that the smaller value is subtracted from the larger value.

$$\text{Percentage difference} = \frac{\text{absolute difference}}{\text{average}} \times 100$$

where

$$\text{average} = \frac{E_1 + E_2 + E_3 + \cdots}{n}$$

E_1, E_2, and E_3 represent the experimental values and n represents the number of experimental values being averaged. Or this can be written as

$$\text{Percentage difference} = \frac{(E_L - E_s)}{\text{average}} \times 100$$

where E_L = largest experimental value and E_s = smallest experimental value.

The metric system is used throughout this laboratory manual. Some of the more common units of length and mass are listed in the accompanying tables.

Length

Unit	Abbreviation	Expression in meters
millimeter	mm	0.001
centimeter	cm	0.01
decimeter	dm	0.1
METER	m	1.0
dekameter	dam	10.0
hectometer	hm	100.0
kilometer	km	1000.0

Mass

Unit	Abbreviation	Expression in grams
milligram	mg	0.001
centigram	cg	0.01
decigram	dg	0.1
GRAM	g	1.0
dekagram	dag	10.0
hectogram	hg	100.0
kilogram	kg	1000.0

LEARNING OBJECTIVES

After completing this experiment, you should be able to do the following:

1. Define the terms accuracy, measurement, least count, precision, and significant figures.
2. Make measurements using a meter stick, beam balance, and electric timer.
3. Calculate the density of a substance when the mass and volume are known.
4. Determine experimentally and theoretically the period of a simple pendulum.
5. Differentiate between percent error and percent difference.

APPARATUS

Meter stick, balance, electric timer, wooden block, simple pendulum.

EXPERIMENT 2   NAME (print) _____ DATE _____
                                  LAST   FIRST
LABORATORY SECTION _____ PARTNER(S) _____

PROCEDURE

1a. The **least count** is the smallest subdivision marked on a measuring instrument; that is, it is the smallest reading that can be made with the instrument without guessing. Determine the least count of the three measuring instruments listed in Data Table 2.1. Record the numerical value of the least count and the unit of measurement. Example: With a meter stick in view note that it is divided and numbered into 100 equal divisions. Each of these numbered divisions is called 1 cm. One centimeter means 1/100 m. Each centimeter is further divided with markings into 10 equal divisions. This is the smallest subdivision on the meter stick. The least count can be expressed as 1/1000 (numerical value) of a meter (unit of measurement), or as 1 (numerical value) millimeter (unit of measurement). However the least count is expressed, it must have the correct unit of measurement for the numerical value stated.

Data Table 2.1   Least Count of Three Measuring Instruments

Measuring instrument	Numerical value	Unit of measurement
Meter stick		
Balance		
Electric timer		

1b. Measure the top of your laboratory table and determine the surface area in square centimeters.

Surface area = _____ × _____ = _____ cm^2

Convert the number of square centimeters to square meters. Show your work.

See Appendix II for conversion factors.

Surface area = _____ meters^2

2. Determine the volume of the wooden block in cubic centimeters. Take three measurements of each dimension of the wooden block and record all significant digits in Data Table 2.2. **Significant digits** are digits read from the measuring instrument plus one doubtful digit estimated by the observer. The estimate will be a fractional part of the least count of the instrument.

    The more precise the measurement, the more exact the information obtained concerning the physical properties of the object that has been measured. **Precision** is an index of the maximum amount a measurement varies from the true value that it represents. A measurement to 0.01 cm is more precise than a measurement to 0.1 cm. The maximum error may be expressed as a plus or minus value. For example, the length of a wood block is 15.4 cm ± 0.1 cm. The maximum error given as ± 0.1 cm means the length is between 15.3 cm and 15.5 cm.

    When using the meter stick, place it on the object to be measured as shown in Fig. 2.1. Never use the end of the meter stick if it can be avoided.

Figure 2.1

Data Table 2.2*

Trial	Length	Width	Height
1			
2			
3			
Average			

* Note: All measurements are to be given in centimeters.

Volume = length × width × height .... _____ cm³

3. Determine the mass of the wood block in grams. This will require the use of the beam balance (see Fig. 2.2), which must be leveled and properly adjusted before an accurate measurement can be made. Ask the instructor for help if needed.

Mass of wood block .... _____ g

Figure 2.2 *Beam balance.*

4. The density of a substance is defined as the ratio of the mass to the volume. Determine the density of the wood block and record below.

   Density = mass/volume.. _____ g/cm^3

5. The second is defined as a fractional part of a mean solar day. Express one second as a fraction of a mean solar day based on our knowledge of there being 60 s in 1 min, 60 min in 1 h, and 24 h in one mean solar day. Show your work.

6. With the use of the electric timer determine your pulse rate. Record the data in Data Table 2.3 and compute the average value.

Data Table 2.3

Trial	Pulses per minute
1	
2	
3	
Average	

7. A simple pendulum consists of a small heavy mass attached to a light string and suspended from a rigid support. The pendulum is set swinging by displacing the mass (called a bob) slightly from its equilibrium position. Use a length of arc (see Fig. 2.3) to pendulum length ratio of 1:10. That is, displace the pendulum bob one-tenth the value of the pendulum length. The word "simple" is used to describe this pendulum because most of the mass is concentrated in the bob. (An ideal simple pendulum would be one in which the entire mass is concentrated at a point. This is impossible to obtain because a bob of any size will have a distribution of mass; also the lightest of strings has mass.) Although the system we are using is not acting as an ideal simple pendulum, a very good approximation of the period can be obtained from the equation $T = 2\pi\sqrt{L/g}$, where the period, T, is defined as the time of one complete swing of the bob. Squaring both sides of this equation, we obtain

$$T^2 = 4\pi^2 \frac{L}{g} \tag{2.1}$$

where T = the period in seconds,

   L = the length of the pendulum measured from the point of support to the center of the bob,

   g = the acceleration due to gravity ($g = 980$ cm/s^2 when L is measured in cm).

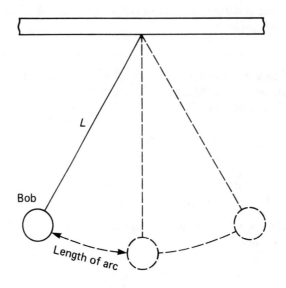

Figure 2.3 *Simple pendulum.*

(a) Determine, experimentally, the period of a simple pendulum with a length of 49 cm. Construct your own data table in the space below and record the data taken. Perform five determinations.

Period (experimental value) .... _____ s

(b) Calculate the theoretical value of the period when L = 49 cm. Use Eq. 2.1.

Period (theoretical value) .... _____ s

How accurate is the above experimental measurement? In theory, the **accuracy** of a measurement refers to how well the experimental value agrees with the true value. However, the true value can never be obtained, and the experimental value is compared with an accepted value known as a standard. How well the experimental value agrees with the standard is usually expressed as a percentage.

$$\text{Percentage error} = \frac{|\text{ Experimental value} - \text{Standard value }|^{*}}{\text{Standard value}} \times 100$$

* Absolute difference is taken. That is, the smaller value is subtracted from the larger value.

(c) Calculate the percentage error using the theoretical value as the standard value. Show your work.

Percentage error...   _____

If no standard value is obtainable, then a percentage difference is determined as follows:

$$\text{Percentage difference} = \frac{\begin{array}{c}\text{absolute difference in}\\\text{experimental values}\end{array}}{\begin{array}{c}\text{average of the}\\\text{experimental values}\end{array}} \times 100$$

(d) Calculate the percentage difference between your largest and smallest experimental value of time. Show your work.

Percentage difference ....   _____

QUESTIONS

1. Differentiate between least count and significant figures.

2. Define measurement and give an example of it.

3. Differentiate between accuracy and precision.

4. Why should the experimenter avoid using the end of a meter stick when making a measurement?

5. Calculate the length in centimeters of a simple pendulum that has a period of one second?

6. Differentiate between percentage error and percentage difference.

7. How can the accuracy of a measurement be increased?

8. How can the precision of the measuring instrument be increased?

The Simple Pendulum

INTRODUCTION

In Experiment 2 the simple pendulum was used to make measurements of length and time to acquire knowledge about the concepts of percent error and percent difference. This experiment will be concerned with how the period of a simple pendulum varies with the length of the pendulum, the mass of the bob, and the magnitude of the bob's displacement from the equilibrium position.

Pendulums are classified as compound or simple. The compound pendulum is one in which the mass of the pendulum is distributed throughout its length. Examples are the pendulum in a grandfather clock and a meter stick held at one end and allowed to swing back and forth. A simple pendulum is one in which the mass of the pendulum is concentrated at a point. It is obvious that the mass of a simple pendulum cannot be located at a point, since a point is defined as position without dimensions. However, the length of the string or wire supporting the mass (called the bob) can be large in comparison to the diameter of the bob. Fig. 3.1 illustrates the parameters of a simple pendulum.

The formula showing the relationship between the parameters of a simple pendulum is:

$$T = 2\pi\sqrt{L/g} \tag{3.1}$$

where T = the period,

L = the length,

g = the acceleration due to gravity = 9.8 m/s^2 = 980 cm/s^2 = 32 ft/s^2.

When both sides of the equation are squared, we obtain

$$T^2 = 4\pi^2 L/g$$

Note: Mass is not in the equation.

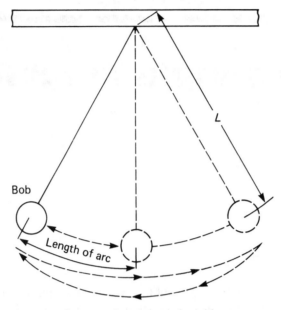

Bob

Length of arc

L

One complete swing = 1 cycle

Figure 3.1 *Simple pendulum*.

LEARNING OBJECTIVES

After completing this experiment, you should be able to do the following:

1. State the difference between a simple and a compound pendulum.
2. Name the parameters that determine the period of a simple pendulum.
3. Determine experimentally the parameter values for a simple pendulum.
4. State how the period of a simple pendulum varies with the (a) length of the pendulum, (b) mass of the bob, and (c) magnitude of the displacement.
5. Calculate the length or period for a simple pendulum when one or the other is given.

APPARATUS

Two different masses for pendulum bobs (hook weights can be used if regular bobs are not available), string, a 2-m stick, timer, balance, and support holder.

PROCEDURE 1

This procedure is for determining the period of a simple pendulum for various lengths.

Step 1 Construct a simple pendulum as shown in Fig. 3.1. Use either given mass (call this mass number one) and make the length of the pendulum for this first step fairly large (160 cm) or whatever is reasonably possible with the means of support you have for the pendulum. The length is the distance from the point of support to the center of the bob.

Displace the pendulum bob approximately 16 cm from its equilibrium position. If you use a length other than 160 cm, use a length-of-arc-to-pendulum-length ratio of 1:10. That is, displace the pendulum bob one-tenth the value of the pendulum length. Determine the time it takes for the pendulum to swing through 10 complete cycles. Before you start your timer, allow the bob to swing through a few cycles. **Caution:** Do not count the first cycle until one cycle has been completed. When the timer is turned on, say "start"; then when the bob comes back to this same position say "one." The best place to start is at one of the end points of the swing where the bob is stopped momentarily. Make three trials and record the data in Data Table 3.1. From the data calculate the period of the pendulum and record its value in the data table.

Step 2 Adjust the length of the pendulum to 80 cm or one-half the length used in Step 1. Complete three trials for the time of 10 cycles and record the data in Data Table 3.1. From the data calculate the period of the pendulum and record this in the data table. Remember the length of arc is to be approximately one-tenth the length of the pendulum.

Data Table 3.1   Mass of pendulum bob: _____ grams

Length of pendulum (cm)	Time in seconds for 10[*] _____ cycles				Period (s)	Period^2 (s)^2
	Trial 1	Trial 2	Trial 3	Average for three trials		
160[*] or _____						
80[*] or _____						
40[*] or _____						
20[*] or _____						

*Cross out this value and insert correct value, if different.

Step 3 Adjust the length of the pendulum to 40 cm or one-fourth the length used in Step 1. Displace the pendulum bob 4 cm or one-tenth the value of the pendulum length you used. Complete three trials for the time of 10 cycles and record the data in Data Table 3.1. **Caution:** Allow the pendulum to swing through enough cycles so that a minimum of 10 s are required. If the time for 10 complete cycles is less than 10 s, then allow the pendulum to swing through 20 complete cycles for the time measurement.

Step 4 Repeat the above procedure with a pendulum length of 20 cm or one-eighth of your original length.

Step 5 Construct a graph with the period of the pendulum in seconds (y axis) versus the length of the pendulum in centimeters (x axis). Refer to Experiment 1 for the correct procedure for constructing a graph.

Step 6 Square all calculated periods and record them in Data Table 3.1. Construct a graph with the period squared (seconds)^2 on the y axis and the length (centimeters) of the pendulum on the x axis. Determine the slope of the curve. Show your work plus the numerical value with units on the graph. Refer to Experiment 1 for an example.

PROCEDURE 2

This procedure is for determining the period of a simple pendulum for a different mass.

Repeat Procedure 1 using mass number two. Make sure the length of the pendulum is measured from the point of support to the center of the bob. Record the data in Data Table 3.2, and compare the periods with those in Data Table 3.1.

PROCEDURE 3

This procedure is for determining the period of a simple pendulum for various displacements of the pendulum bob.

Step 1 Construct a simple pendulum as shown in Fig. 3.1. Use mass number two and make the length of the pendulum 160 cm or the same length you used in Procedure 1 Step 1. Displace the pendulum bob 16 cm or one-tenth the pendulum length you used. Make three trials and record the data in Data Table 3.3. From the data calculate the period of the pendulum and record in the data table. *Note:* For this length of pendulum the period should be the same as you recorded in Data Tables 3.1 and 3.2.

Step 2 Using the same length and mass as in Step 1, displace the pendulum bob 32 cm from its equilibrium position or one-fifth the pendulum length you used. Make three trials and record the data in Data Table 3.3. From the data calculate the period of the pendulum and record in the data table.

Step 3 Repeat Step 2 displacing the pendulum bob 48 cm from its equilibrium position or three-tenths the pendulum length you used. Record all data in Data Table 3.3.

Step 4 Repeat Step 2 displacing the pendulum bob 64 cm or four-tenths the pendulum length you used. Record all data in Data Table 3.3.

LAST   FIRST

Data Table 3.2   Mass of pendulum bob _____ grams

Length of pendulum (cm)	Time in seconds for 10* _____ cycles				Period (s)	Period² (s)²
	Trial 1	Trial 2	Trial 3	Average for three trials		
160* or _____						
80* or _____						
40* or _____						
20* or _____						

*Cross out this value and insert correct value, if different.

Data Table 3.3   Length of pendulum _____ cm; mass of bob _____ grams

Length of arc (cm)	Time in seconds for 10* _____ cycles				Period (time of one cycle) in s
	Trial 1	Trial 2	Trial 3	Average for three trials	
16* or _____					
32* or _____ _____					
48* or _____ _____					
64* or _____ _____					

*Cross out this value and insert correct value, if different.

QUESTIONS

1. Why were you asked to complete more cycles in Procedure 1 Step 3, if the total time was less than 10 s for 10 cycles?

2. What type of graph (straight line, parabola, hyperbola) was obtained in Procedure 1 Step 5?

3. Which variable (period or length) does the graph show to be increasing the fastest?

4. What type of graph was obtained in Procedure 1 Step 6?

5. State how the period of the simple pendulum varied experimentally with the (a) length of the pendulum, (b) length of arc (displacement of the bob), and (c) mass of the bob.

6. A simple pendulum has a length L and a period T. If the length is made $4L$, what will be the new period? Refer to data tables for the answer.

7. Calculate the period of your pendulum for a length of 160 cm or for the maximum value you used. How does this compare with your experimental value? Calculate the percent error. Use the calculated value as the standard value.

8. Assuming the length remains the same, would the period of a simple pendulum be greater or less on the surface of the moon? (*Note:* The acceleration (g) due to gravity on the moon is one-sixth of that on Earth.) Explain your answer.

Uniform and Accelerated Motion

INTRODUCTION

An object that is undergoing a change in position is said to be moving. The motion may be either simple straight-line motion, where the change in distance is the same for each period of time, or it may be very complex, as in a tumbling satellite orbiting the Earth. Our study in this experiment deals with simple straight-line motion with uniform change in the distance, and straight-line motion with uniform change in the velocity. A good example of straight-line motion is the movement of a glider along a linear air track. A close observation of the glider will reveal that its position changes at a uniform rate; that is, the distance traveled is the same for equal periods of time. We can say it another way—the distance traveled is directly proportional to the time. This can be written as

$$\text{distance } (d) \propto \text{time } (t)$$

or
$$d = \overline{v}t$$

where $\overline{v}$ (pronounced v bar) is a proportionality constant. Solving for $\overline{v}$, we obtain

$$\overline{v} = \frac{d}{t}$$

The quantity $\overline{v}$ is known as the average velocity of the glider.

An object whose velocity is changing is said to be accelerating. *The velocity may be increasing, decreasing, or changing direction. Any of these causes a change in the velocity.* In this experiment we will be dealing with a change in velocity due to an increase. In observing the glider traveling down the inclined plane, you will discover that the velocity change will be the same for equal periods of time. This means that the velocity change is directly proportional to the change in time. This can be written as

$$\text{change in velocity } (\Delta v) \propto \text{change in time } (\Delta t)$$

or
$$\Delta v = a \, \Delta t$$

where a (acceleration) is a proportionality constant. Solving the equation for a we obtain

$$a = \frac{\Delta v}{\Delta t}$$

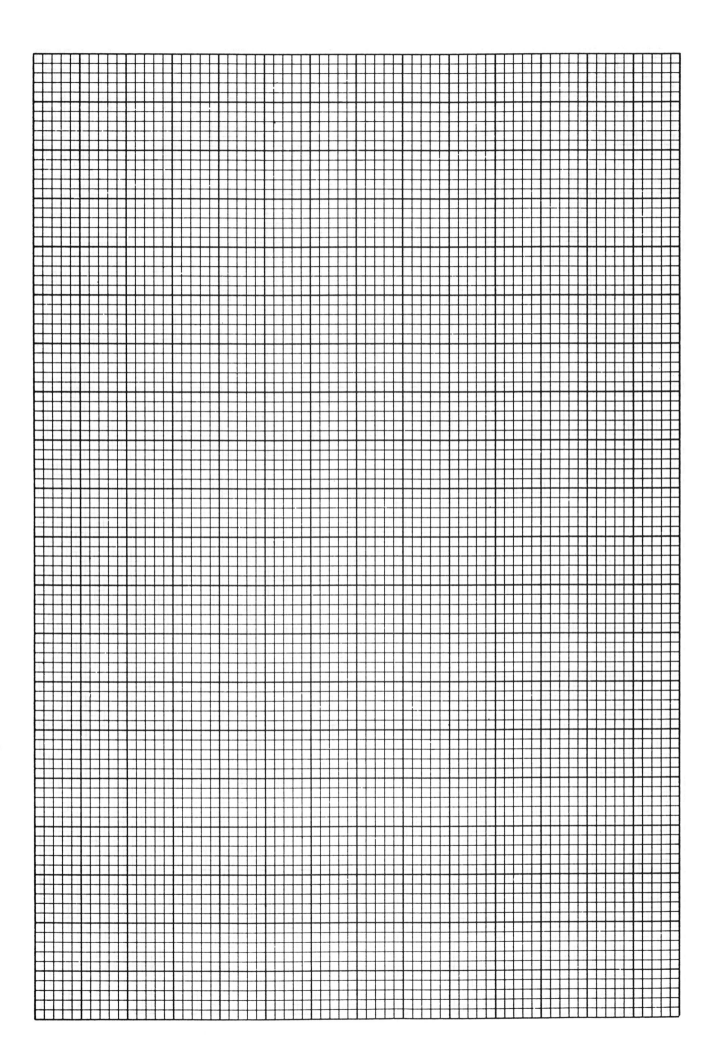

Determining g, the Acceleration of Gravity

INTRODUCTION

A glider sliding down an inclined air track or a steel ball rolling down a V-grooved 2 in × 6 in wooden plank has the force of gravity pulling vertically downward on it. The force acting is equal to the weight of the object. A second force is also acting on the object. The second force is exerted by the supporting air track or the wood plank, which acts perpendicular to the surface of the supporting air track or wood plank and is called the normal force. Fig. 5.1 shows the two forces and the resultant of the two forces that act parallel to the incline to produce an acceleration of the object down the incline. An analysis of the two forces shows that the resultant force F down the incline is equal to $mg \sin \theta$. (See Fig. 5.2.) The sine of the angle θ is equal by definition to the side opposite the angle divided by the hypotenuse. Thus, the height of the incline divided by the length of the incline equals $\sin \theta$, or $\sin \theta = h/L$. This force of $mg \sin \theta$ or mgh/L is the unbalanced force acting to accelerate the mass down the incline. Newton's second law of motion states the relationship between the unbalanced force, mass, and acceleration of the object. This can be written as

$$a = F/m$$

Since $F = mgh/L$,

$$a = \frac{mgh/L}{m} = gh/L$$

or

$$g = \frac{aL}{h} \tag{5.1}$$

To determine the value of g we need to measure the values of h and L and determine the value of a. The acceleration a can be determined by the following relationship:

$$a = \frac{v_f - v_o}{t} \text{ by definition}$$

If $v_o = 0$, then $a = \dfrac{v_f}{t}$. Since $v = \dfrac{(v_f + v_o)}{2}$, if $v_o = 0$, then $v = \dfrac{v_f}{2}$ or $v_f = 2\overline{v}$.

But $\overline{v} = \dfrac{d}{t}$ by definition. Therefore

$$a = \frac{2v}{t} = \frac{2d/t}{t} \tag{5.2}$$

where d is the distance traveled and t is the time for the object to travel the distance d.

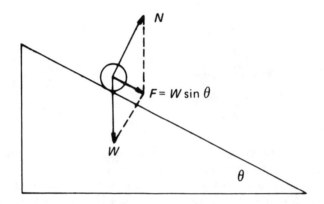

Figure 5.1 *Diagram illustrating the two forces acting on an object traveling down a frictionless inclined plane. Air resistance is neglected. W represents the weight of the object, N represents the normal or perpendicular force on the object, and F is the resultant of W and N.*

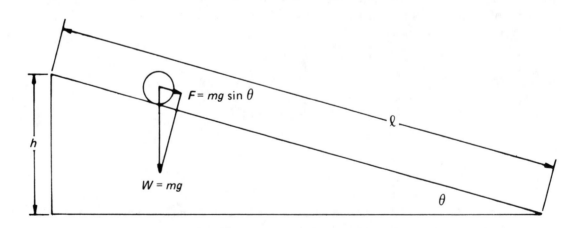

Figure 5.2 *Steel ball rolling down an incline due to the pull of gravity, or the weight of the ball.*

LEARNING OBJECTIVES

After completing this experiment, you should be able to do the following:

1. Determine experimentally the acceleration (g) of gravity at the surface of the Earth.
2. Calculate the percent error of your measurement with the accepted value of g at your surface location.

APPARATUS

Linear air track, glider, meter stick, five timing devices (stop watches or electric timers), and masking tape. If the air track is not available, then a 2 in × 6 in wood plank 10 ft long plus a 3/4-in-diameter steel ball can be used. The wood plank must have a trough milled into one edge of the plank. A V-grooved trough is preferred.

PROCEDURE

1. If the air track is to be used, ask the instructor to adjust the air track at an angle with the table top. This can be done by placing a small block (about 2 cm high) under one end of the air track. If the 2 in × 6 in plank is used, ask the instructor to elevate the plank at an angle of about 5°. Use small pointed pieces of masking tape and mark off five equal distances (50 cm) along the incline. Place the glider or the steel ball on the incline at the piece of tape farthest up the incline. This is the starting point. Release the object and start all timers simultaneously. Determine the time for the moving object to travel the distance from rest at the starting point to each marker. Make three trials and record the values obtained for each distance in Data Table 5.1.

2. Measure the height h and the length L of the incline and record in the data table.

CALCULATIONS

1. Plot a graph with distance on the vertical axis and time squared on the horizontal axis. Make sure the graph contains all the information concerning graphs as given in Experiment 1.

2. Calculate the acceleration of the object down the incline. Use Eq. 5.2. Show your work. The final answer will be the average acceleration for the three distances.

_____

Data Table 5.1

Distance traveled (d)	50 cm	100 cm	150 cm	200 cm	250 cm
Time to travel distance (d), in seconds					
Average time (t), in seconds					
Height of the incline _____ cm					
Length of the incline _____ cm					

3. Calculate the experimental value for the acceleration of gravity. Use the value of (a) from 2 above.

$$g = \frac{aL}{h}$$

_____

4. Calculate the percent error using 980 cm/s^2 as the correct value of g. Show your work.

_____

QUESTIONS

1. What was your greatest source of error in this experiment? Justify your answer.

2. What kinematic concept of the moving object does the slope of the graph give?

3. If the initial velocity was not zero, how would the value of g compare with the value you obtained? For instance, suppose you accidentally gave the object a slight shove down the incline at the starting point.

4. What will be the value of the acceleration when θ (theta) in Fig. 5.2 is increased to 90°?

5. Is the acceleration due to gravity always pointing vertically downward even for an object whose velocity is vertically upward? Explain your answer.

45

Newton's Second Law

INTRODUCTION

A mass whose velocity is changing at a constant rate is said to be undergoing uniform acceleration. The velocity may be increasing; if so, the acceleration is considered to be positive. Or the velocity may be decreasing; then the acceleration has a negative value. In either case, the change in velocity must be the same for each period of time for uniform acceleration.

Newton's second law of motion gives the relationship between mass, acceleration, and the unbalanced force causing the mass to accelerate:

$$a \propto \frac{F}{m} \text{ or } a = k\frac{F}{m}$$

where a = acceleration of total mass in direction of the unbalanced force,

   F = the unbalanced force acting to produce acceleration,

   m = total mass accelerated,

   k = proportionality constant.

The value of k depends on the units of F, a, and m. In this experiment we shall express F in dynes, a in centimeters per second squared, and m in grams. This will give k a value of 1; therefore, it may be dropped from the equation. Remember, to change mass units to force units, multiply the mass by g, the acceleration due to gravity.

To study the relationships among these quantities we shall use a single pulley that has very little friction, and masses supported as shown in Fig. 6.1. The mass of the pulley and the mass of the string will be neglected in performing the experiment.

LEARNING OBJECTIVES

After completing this experiment, you should be able to do the following:

1. State Newton's second law of motion both in words and in symbol notation.
2. Determine experimentally the relationship between the concepts mass, acceleration, and the unbalanced force causing the mass to accelerate.

Figure 6.1

APPARATUS

Precision ball-bearing pulley, masses, cord, electric timer, support rods, clamps, and 2-m meter stick.

LAST   FIRST

LABORATORY SECTION _____ PARTNER(S) _____

PROCEDURE

1. Assemble the apparatus as shown in Fig. 6.1 if this has not already been done by a previous laboratory instructor. It is important that the pulley be positioned to turn freely in a vertical plane. Otherwise, there will be an opposing force that will give poor results for the experiment. Keep the unbalanced force constant and determine experimentally how the acceleration varies as the total mass is varied. Perform two trials using two different masses. Enter data in Data Table 6.1.

   To obtain the data, proceed as follows: With one mass resting on the floor, and a suitable rider or riders (m_3) on the top mass, simultaneously release the system and start the timer. The rider or riders should not be more than 3% of the total mass. When the bottom of the upper mass reaches the floor, stop the timer. Make six independent determinations of this time and take the average. Distance d is measured from the bottom of the upper mass to the floor. See Fig. 6.1.

2. Keep the total mass constant* and determine how the acceleration varies with the variation of the unbalanced force. Perform two trials using two different rider values. Enter data in Data Table 6.2.

Data Table 6.1

			Trial 1 $m_1 = m_2 = x$ grams	Trial 2 $m_1 = m_2 = y$ grams
Distance d the mass travels, in centimeters				
Unbalanced force F, in dynes (constant for all trials) $F = mg = [(m_2 + m_3) - m_1]g$				
Total mass m moved, in grams (different for each trial)				
Time to travel distance d, in seconds	Three runs with rider on left	1		
		2		
		3		
	Three runs with rider on right	1		
		2		
		3		
	Average			

m_3 = rider or riders

_____

* *Note:* To keep the total mass constant the experimenter should change some of the riders from one side of the pulley to the other.

CALCULATIONS

Compute the acceleration, first from the measurements taken of the distance traveled and the time; second, by using Newton's second law. Show your work.

1. $a = \dfrac{2d}{t^2}$

2. $a = \dfrac{[(m_2 + m_3) - m_1]g}{m_1 + m_2 + m_3}$

	Trial 1	Trial 2
Acceleration from measurements of distance and time		
Acceleration from Newton's second law		
Percent difference		

Data Table 6.2

			Trial 1 $m_1 = m_2 = x$ grams	Trial 2 $m_1 = m_2 = y$ grams
Distance d the mass travels, in centimeters				
Unbalanced force F, in dynes (different for each trial) $F = mg$				
Total mass m moved, in grams (constant for all trials)				
Time to travel distance d, in seconds	Three runs with rider on left	1		
		2		
		3		
	Three runs with rider on right	1		
		2		
		3		
	Average			

*m = the difference between all masses on each side of the pulley.

CALCULATIONS

Compute the acceleration (1) from the measurements taken of the distance traveled and the time and (2) by using Newton's second law. Show your work and record the answers in the space provided.

	Trial 1	Trial 2
Acceleration from measurements of distance and time		
Acceleration from Newton's second law		
Percent difference		

QUESTIONS

1. What have the data and calculations shown concerning the acceleration–first, as a function of the mass when the unbalanced mass is held constant, and second, as a function of the unbalanced force when the total mass is held constant? Answer this by completing the following two statements:

   (a) When the total mass that is accelerating increases (the unbalanced force held constant), the acceleration will _____.

   (b) When the unbalanced force increases (the total mass held constant), the acceleration will

   _____.

2. Distinguish between mass and weight.

3. Should the mass of the string be added to the total mass being moved by the unbalanced force? Why or why not?

4. Since it is the addition of m_3 to the system that produces the acceleration of the system, why is the unbalanced force in Procedure 1 equal to $[(m_2 + m_3) - m_1]g$ and not simply $m_3 g$?

5. Where is the greatest possible source of error in this experiment? Explain your answer.

Hooke's Law for a Vibrating Spring

INTRODUCTION

A common type of vibrating system is a spring with a heavy mass (m) suspended from one end. If the mass is set in an up-and-down (vertical) motion, the repeated stretching and compression of the spring follows certain rules of periodicity that demonstrate some of the characteristics of wave motion.

The period of the vibration (time for one complete up and down oscillation) may be defined as

$$T = 2\pi\sqrt{M/k}$$

where T is the period in seconds,

         π is equal to 3.14,

         M is the sum of the attached mass plus one-third of the mass of the spring itself [$M = m + 1/3$(spring mass)] in grams,

and k is the spring constant in dynes/cm.

The same spring constant (k) is also used in Hooke's law. This law applies to the stretching of a spring when a mass is hung on the end of it and gravity pulls on the mass to produce a stretching force. The proportionality between force (F) and stretch (X) is known as Hooke's law. This relationship can be written in equation form as

$$F = kX$$

where F is the applied force mg in dynes,

         X is the distance the spring is stretched, in cm,

and k is the same spring constant as above, in dynes/cm.

LEARNING OBJECTIVES

After completing this experiment, you should be able to do the following:

1. State and explain Hooke's law.
2. Write down and explain the equation for the period of an oscillating mass on a spring.
3. Determine the spring constant for any spring using a set of weights and a meter stick.
4. Calculate the period of oscillation for a mass on a spring both theoretically and experimentally.
5. Access the accuracy of your period calculations using a percentage error calculation.

APPARATUS

Spring, set of calibrated masses, triple-beam balance or spring balance to weigh spring, support stand, clamp.

PROCEDURE

First we must measure the spring constant of the particular spring you are going to use in this experiment. Attach the spring to the support clamp and *carefully* hang enough mass on the spring to stretch it about 2.0 to 3.0 cm. This mass will be referred to as the *incremental mass* throughout the rest of the experiment. Record the value of the incremental mass in the space provided below. Remove incremental mass from the spring.

Incremental mass = _____ g

Now adjust the support clamp so that the bottom end of the spring is about 30 cm above the top of the table. Carefully measure the distance from the tabletop to some identifiable point near the lower end of the spring. This distance is h_o. (See Fig. 7.1.) Record this value in the space provided at the bottom of this page.

Again hang the same incremental mass on the spring that you used in your earlier calibration of the spring. The spring should stretch about 2.0 to 3.0 cm, as it did before. Carefully measure the distance from your reference point on the spring to the top of the table. This is the distance h for Trial 1. Record this value in Data Table 7.1 along with the value of the mass hanging on the spring. This should be the same as the incremental mass that you recorded in a previous step.

Repeat the above procedure for values of mass equal to 2, 3, 4, 5, and 6 times the incremental mass. The spring should stretch an additional 2.0 to 3.0 cm for each trial. Carefully measure the height h above the table of your reference point on the spring for each trial. Record the height, h, and the mass, m, for each trial in Data Table 7.1.

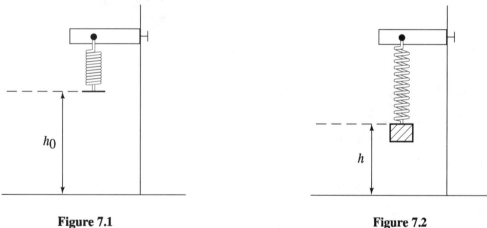

Figure 7.1   Figure 7.2

The force of gravity as it acts on the mass hung from the spring can be found by multiplying this mass (in grams) times the acceleration of gravity, 980 cm/s^2. Record the force pulling downward on the spring in each of your six trials in Data Table 7.1.

Starting height h_o = _____ cm

Data Table 7.1

Trial	Total mass on spring m (grams)	Height above the table h (cm)	$X = h_0 - h$ (cm)	Force $F = m \times 980 \text{ cm/s}^2$ (dynes)	$k = \dfrac{F}{m}$ (dynes/cm)
1					
2					
3					
4					
5					
6					
				Average value of k for all six trials	

Calculate the spring constant k using Hooke's law for each of the six trials recorded in Data Table 7.1.

Weigh the entire spring and find the spring mass.

Spring mass = _____ g

The effective mass of the vibrating spring would now be:

m = hanging mass

$M = m$ + one-third of spring mass = _____ g + 1/3(_____) g

$M =$ _____ g

Now we can calculate the theoretical period at which the spring and mass will oscillate if they are set in motion. Using the data from Data Table 7.1, determine the theoretical period for the six vibrating spring-and-mass combinations and record your values for T in Data Table 7.2.

$$T = 2\pi \sqrt{M/k}$$ (Use m and k from Data Table 7.1 above to calculate M and then T.)

To see how well this calculation agrees with the actual time for one oscillation, set the spring-and-mass system oscillating and measure the time for 20 complete vibrations with a time clock or stop watch. Use the same mass on the spring that you used in Trial 1 in Data Table 7.1. Repeat this procedure for the other five trial masses in Data Table 7.1 and record the times for 20 oscillations for each mass value in Data Table 7.2.

To calculate the experimental period for each trial, divide your measured time for each trial by 20 and fill in this data in Data Table 7.2.

$$T = \frac{t}{20}$$

How does your experimental value for T compare with the theoretical value found by calculation? To determine this, do a percentage error calculation for each trial and see.

Data Table 7.2

Trial	Effective mass (M) of spring and hanging mass (grams)	Theoretical period (T) $T = 2\pi \sqrt{M/r}$ (s)	Measured time (t) for 20 oscillations (s)	Experimental period—time for 1 vibration (s)	Percentage error in experimental period compared to theoretical period
1					
2					
3					
4					
5					
6					

QUESTIONS

1. Why must you use the same reference point on the spring for each height measurement?

2. Why did we use the average value for k from Data Table 7.1 rather than one of the individual trial values?

3. Do a calculation of the theoretical value of the period for the oscillating mass for Trial 4 *without* correcting the effective mass by adding one-third of the mass of the spring. Do you agree that the calculation must be corrected in this way for good results?

Centripetal Acceleration and Force

INTRODUCTION

A mass or particle moving in a circle with constant speed is said to be in **uniform circular motion**. Although the speed of the mass (or particle) is constant, its velocity is not constant because the direction of motion is continually changing. Because the velocity is continually changing, the mass is undergoing continual acceleration. However, the direction of travel of the mass is changing. Hence, the acceleration must be directed at right angles to the direction in which the body is moving (otherwise its speed would increase); that is, the acceleration is directed toward the center of the circle. The force causing the acceleration, called the **centripetal force**, may arise from mechanical or electrical phenomena. In this experiment, a spring (mechanical phenomenon) will supply the centripetal force. By Newton's third law, there will be an equal and opposite force acting on the spring. This force is called the **centrifugal force**. Note the difference between the two forces. (See Fig. 8.1.)

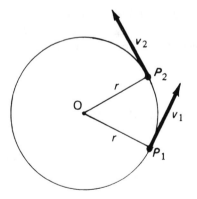

Figure 8.1 *Velocity of a particle moving in uniform circular motion.*

The acceleration of a mass moving in uniform circular motion is given by

$$a = \frac{v^2}{r} \tag{8.1}$$

where r is the radius of the circle in which the mass is traveling and v is the linear velocity of the mass. According to Newton's second law, the centripetal force F_c producing this acceleration is

$$F_c = ma = \frac{mv^2}{r} \tag{8.2}$$

It is convenient to express v in terms of other quantities. The circumference of the circle, d, which the mass draws out as it travels is simply

$$d = 2\pi r$$

where r is the radius of the circle. If the mass makes n revolutions in a time t, the total distance d traveled by the mass in this time is

$$d = nd = 2\pi rn$$

and the velocity of the particle is simply

$$v = \frac{d}{t} = \frac{2\pi rn}{t}$$

so that Eq. 8.2 may now be written

$$F_c = \frac{mv^2}{r} = \frac{m}{r}\left(\frac{2\pi rn}{t}\right)^2 = 4\pi^2 mr \left(\frac{n}{t}\right)^2 \tag{8.2a}$$

By use of Eq. 8.2a, it is now possible to determine the centripetal force on a rotating body by measuring the mass, the radius of rotation, the number of revolutions, and the total time for the revolutions.

LEARNING OBJECTIVES

After completing this experiment, you should be able to do the following:

1. Define uniform circular motion.
2. Determine experimentally the centripetal force on a rotating object by measuring the mass, the radius of rotation, the number of revolutions, and the time for the revolutions.
3. Compare the observed and calculated values of the centripetal force.

APPARATUS

Centripetal force apparatus; stopwatch (or clock with sweep second hand); weight hanger; assorted slotted masses; rods and clamps.

PROCEDURE

1. This experiment makes use of the simple apparatus shown in Fig. 8.2. The apparatus is essentially a vertical rod with a crossbar and a measuring stick attached to another vertical rod. The rod and crossbar assembly can be rotated easily by pulling a string wound around the vertical rod.

   Hang the known mass from the rotating crossbar so that it is a few centimeters from the end of the unstretched spring and its tip is just above the measuring stick. Record the measurement to the point just below the tip of the mass. Place an identifying mark at this point. This measurement is made from the center of the vertical revolving rod. It is r in Eq. 8.2a.

Figure 8.2   *Centripetal force apparatus.*

2. Attach the mass to the spring (it will be drawn inward). Wind the string about the vertical rod and pull smoothly until the mass (1) moves with uniform velocity and (2) consistently passes, vertically, over the mark. Practice this two or three times. It is important that conditions 1 and 2 both be satisfied in this experiment. Count the number of revolutions n and observe the time t from that time when the mass passes over the mark in the first revolution to the time when the mass passes over the mark in the last revolution. The latter will occur approximately when the string is completely unwound. Compute the ratio n/t and use Eq. 8.2a to compute F_c.

3. While the mass is still attached to the spring, pull the mass out until it is over the mark. Measure the distance from the center of the revolving rod to the end of the spring nearest the mass. Call this R_1. Remove the mass and allow the spring to rest on the measurement stick. Again measure the distance to the same end of the spring. Call this distance R_2. The difference $R = R_1 - R_2$ is the amount the spring was stretched.

4. Place the spring on the crossbar provided and note its position (Fig. 8.3a). Attach a weight hanger and weights until the spring stretches a distance R (Fig. 8.3b). Record the mass (including that of the hanger). This mass times the acceleration of gravity is the force F_s necessary to stretch the spring R cm. Compare this with the result obtained from Eq. 8.2a.

5. Make three trials for n and t. Repeat Procedures 1 through 4 for a different mass, keeping r constant.

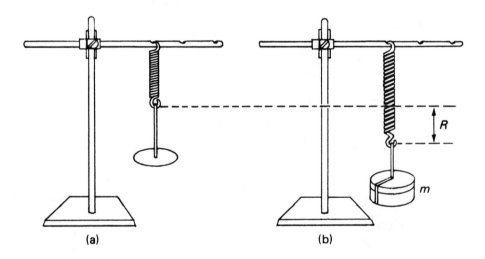

(a)

(b)

Figure 8.3 *Spring and masses. Position (a) shows spring without mass. Position (b) shows spring with mass attached.*

Data Table 8.1

Trial	n	t	r	$R = R_1 - R_2$	F_c	Total mass	F_s	Percent difference
1								
2								
3								

Data Table 8.2

Trial	n	t	r	$R = R_1 - R_2$	F_c	Total mass	F_s	Percent difference
1								
2								
3								

QUESTIONS

1. Assume that the mass was not supported by the string while in motion. If the spring were suddenly unhooked from the mass, what would happen to the mass?

2. How did F_c change when you changed masses? Did you expect this? Why?

Laws of Equilibrium

INTRODUCTION

An object is in equilibrium when the summation of all forces and torques acting on the object equals zero, or

$$\Sigma F = 0 \quad \text{and} \quad \Sigma L = 0$$

where Σ = summation,

   F = force,

   L = torque.

A state of equilibrium does not mean that the object is at rest. The object may be at rest or it may be in motion.

LEARNING OBJECTIVES

After completing this experiment, you should be able to do the following:

1. Define the terms torque, center of gravity, theoretical mechanical advantage (TMA), actual mechanical advantage (AMA), and efficiency.
2. State in words and symbols the two laws of equilibrium.
3. Determine experimentally the TMA, AMA, and efficiency of some simple machines.

APPARATUS

Moment-of-force apparatus (uniform meter stick, three knife-edge clamps, support for meter stick, set of hooked weights) inclined plane, Hall's carriage, weight hanger, set of slotted weights, one unknown weight, two single pulleys, two double pulleys, cord, table supports.

PROCEDURE

1. The center of gravity of an object is defined as that point at which all its mass could be concentrated to give the same resultant force as all individual forces acting on each particle making up the object. Determine the center of gravity of the meter stick using the knife-edge support and stand. See Fig. 9.1. Remove weight hangers.

   Center of gravity of the meter stick ....   _____

Figure 9.1   *Moment-of-force apparatus.*

2. A torque, which tends to produce a rotation, is defined as a force times a distance. The distance is measured from the axis of rotation to the point of the applied force. This distance is called the lever arm. The force is the component acting perpendicular to the lever arm.

   (a) Balance the meter stick on the support stand. Place a 50-g mass at the 20-cm point of the meter stick and determine, first theoretically and then experimentally, where a 100-g mass should be positioned to place the system in equilibrium. See Fig. 9.2. Show your calculations. Use center of gravity of meter stick for axis of rotation.

clockwise moments = counterclockwise moments

Figure 9.2

4. (a) Assemble a single pulley as shown in Fig. 9.4. Record in Data Table 9.2 the value of the force F needed to place the system in equilibrium.

Figure 9.4

(b) Set up the pulley system as shown in Fig. 9.5. Record the value of the force F needed to place the system in equilibrium. *Add weight of lower pulley to obtain total load.*

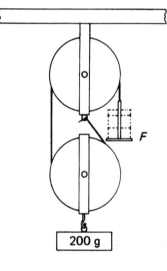

Figure 9.5

(c) Assemble the pulley system as shown in Fig. 9.6. Record the value of the force F needed to place the system in equilibrium. Add weight of movable pulleys to obtain total load.

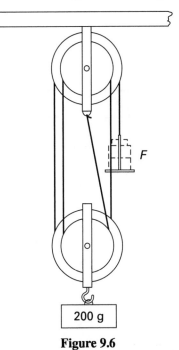

200 g

Figure 9.6

(d) Assemble the pulley system as shown in Fig. 9.7. Calculate the value of the force F needed to place the system in equilibrium, then determine F experimentally. Record the data and make the calculations called for in Data Table 9.2.

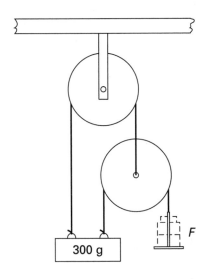

300 g

Figure 9.7

Data Table 9.2

Load, w, in dynes	Mass (g) added	Force F, in dynes	L	h	TMA	AMA	Efficiency
(a)							
(b)							
(c)							
(d)							

Note: L is distance *F* moves, *h* is distance load moves.

QUESTIONS

1. Distinguish between force and torque.

2. State in words the two laws of equilibrium.

3. Distinguish between TMA and AMA.

4. Is the TMA always greater than the AMA for a machine? Explain.

5. Define efficiency. Give an example.

Experiment 10

Waves

INTRODUCTION

Waves are a means of propagating energy from particle disturbances. Most students are familiar with water waves, sound waves, light waves, and waves in stretched strings like piano or guitar strings. Particle disturbance means that the particle has been displaced from its equilibrium position by some agent. Once displaced, the particle tends to return to its original position. In the process, a wave will be propagated outward from the disturbance, transferring energy. The propagation of energy from the disturbance is called **wave motion.**

Waves are classified as transverse or longitudinal. **Transverse waves** have the particle displacement perpendicular to the motion of the wave. **Longitudinal waves** are those in which the particle displacement is in the same direction as the wave motion.

All waves have certain fundamental properties. A few of these properties are velocity, wavelength, period, frequency, and amplitude. These are illustrated in Fig. 10.1 on the next page.

The **velocity** of the wave is the distance the wave travels per unit of time, and in this experiment it is measured in centimeters per second. The **wavelength** is the distance between two similar points on any two consecutive waves measured in centimeters or some other unit of length. The **period** of a wave is the time required for one wavelength to pass any point along the direction of travel. The **frequency** is the number of vibrations or cycles per unit time the particle disturbance is taking place. Frequency is usually measured in cycles per second, or Hertz. The period is the reciprocal of the frequency. You will recall that the reciprocal of a number is one divided by the number. The relation here, therefore, would be written as

$$T = \frac{1}{f}$$

where T = period in seconds and f = frequency in cycles per second.

The relation among velocity, frequency, and wavelength is shown by the equation

$$v = \lambda f$$

where v = velocity,

λ = wavelength,

f = frequency.

Figure 10.1

LEARNING OBJECTIVES

After completing this experiment, you should be able to do the following:

1. Define the terms wave, wavelength, frequency, period, and amplitude of a wave.
2. Differentiate between longitudinal and transverse waves.
3. Demonstrate the transfer of energy by waves.
4. Measure the wave velocity in a stretched rubber rope.
5. Measure the wavelength of a sound wave.

APPARATUS

Metal can with plastic lid (large-size coffee can), candle, 2-m meter stick, rubber cord, timer, coiled spring, tuning forks (use frequencies high enough to give two or three maximum and minimum points for air column in closed pipe), rubber hammer, resonance apparatus, large water tray, cork stopper.

PROCEDURE

1. Assemble the metal can, meter stick, and candle as shown in Fig. 10.2. Determine if you are able to extinguish the candle flame 1 m (or less) from the can by lightly thumping the plastic lid with a rubber hammer. Explain your results.

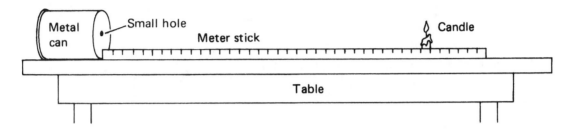

Figure 10.2

2. The velocity of a wave in a stretched string or rubber cord is given by the following:

$$v = \sqrt{\dfrac{F}{m/L}}$$

where v =   velocity,

        F =   tension in cord measured in units of force,

   m/L =   mass per unit length.

Attach the rubber cord to one wall of the laboratory, then apply tension to stretch the cord across the full length of the room. Displace the cord as shown in Fig. 10.3. Observe the wave travel the length of the cord, be reflected at the wall, and return to your hand.

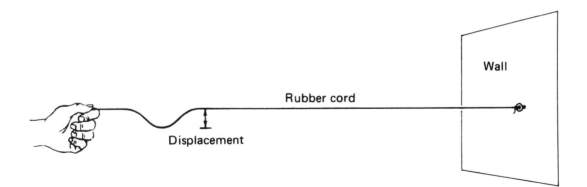

Figure 10.3

With the electric timer determine the time (three trials) it takes a wave to travel the total distance. Measure the distance traveled and calculate the velocity of the wave.

$$v = \frac{s}{t} = \underline{\hspace{4cm}}$$

Repeat the experiment with increased tension in the cord.

$$v = \frac{s}{t} = \underline{\hspace{4cm}}$$

What type of wave is traveling in the cord? Explain your answer.

3. Stretch the coiled spring along the floor. Displace the spring at one end by compressing a few coils and releasing. Observe the wave motion along the spring. Explain the type of wave observed.

4. Fill the large tray with water, then place it on the laboratory table. Place the cork stopper in the water near one end of the long dimension of the tray as shown in Fig. 10.4. Disturb the water near the other end of the tray by dropping a coin flat into the water.

Figure 10.4

Observe the wave motion on the water and the movement of the cork. What type of wave are you observing? Explain your answer.

5. In the introduction, wavelength was defined and illustrated. Fig. 10.5 illustrates the component sections of one complete wavelength.

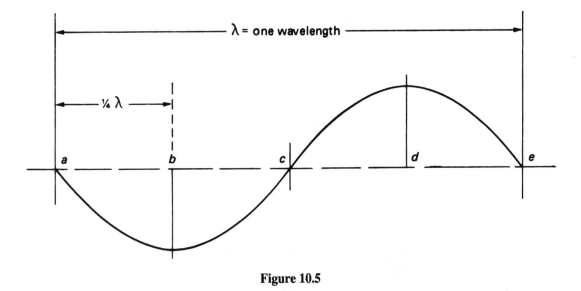

Figure 10.5

Note that minimum particle displacement occurs at positions *a*, *c*, and *e*, and maximum displacement at *b* and *d*. The distance *a-b* is one-fourth wavelength. Likewise the distances *b-c*, *c-d*, and *d-e* are one-fourth wavelength. Observe that one maximum and one minimum value occurs in each quarter-wavelength. Fig. 10.6 illustrates the particle displacement along a quarter-wave tube containing a vibrating column of air produced by a tuning fork. The air at the open end of the tube will vibrate at the maximum value because at the open end the air is free to move. At the lower end of the tube, minimum particle displacement will take place because the tube is closed and particle vibration is restricted.

Figure 10.6 *Air column in closed pipe.*

Using such a resonance apparatus, we can determine the wavelength of a sound wave. A simple model of this apparatus is shown in Fig. 10.7.

The height h may be a quarter-wavelength or a whole-odd-number multiple of a quarter-wavelength ($\lambda/4$, $3\lambda/4$, $5\lambda/4$, . . .). In using the apparatus to determine the wavelength of sound, all that is necessary is to adjust the length of the air column by adjusting the water level for the loudest sound heard. Record the air-column length for this maximum loudness, then adjust the level again for the next maximum level. The distance between two maximums is one-half wavelength.

The velocity of sound in air in centimeters per second can be calculated from the following equation:

$$v = 33{,}200 + 60\,T$$

78

Figure 10.7 *Resonance apparatus.*

where v = velocity of sound in air,

    33,200   = velocity of sound in air at 0° Celsius, in cm/s,

         T = temperature of air in degrees Celsius.

Using this information and the equation relating velocity, wavelength, and frequency, $\lambda = v/f$, calculate the value of a quarter-wavelength for the 1024-cycles-per-second tuning fork.

Calculated value of half-wavelength .... _____ cm

Strike the 1024-cps tuning fork with the rubber hammer. Place the vibrating fork at the top of the air column and adjust the water level until the air column is vibrating at its maximum loudness. You will be able to ascertain the maximum loudness by listening. Record the length of the air column. Readjust the water level to the next maximum loudness position. Record the length of the air column and calculate the value of a quarter-wavelength.

Experimental value of half-wavelength .... _____ cm

Percent difference .... _____ %

Determine the wavelength and frequency of an unknown tuning fork using the resonance apparatus.

Wavelength .... _____ cm

Frequency... _____ cps

Ask the laboratory instructor for the correct value of unknown frequency, and determine the percent error.

Percent error... _____ %

QUESTIONS

1. Define wave motion and give an example.

2. Distinguish between longitudinal and transverse waves.

3. Define the velocity, wavelength, and frequency of a wave and give an example of each. Use symbol notation in the definitions.

4. State the relationship between velocity, wavelength, and frequency of a wave.

5. Define the amplitude of a wave and give an example.

6. State the relationship between the frequency and the period of a wave.

7. How does the velocity of a wave in a rubber cord vary with the tension in the cord?

8. Why is there always a maximum value for the vibration of the air at the open end of a resonance tube?

9. How does the velocity of sound in air vary with the air temperature?

Interference of Light Waves

INTRODUCTION

When matter is disturbed, energy radiates from the disturbance; this radiation of energy is called **wave motion**. Wave motion has physical properties such as velocity, frequency, period, wavelength, and amplitude as defined and explained in Experiment 10. Another property of a wave is phase. The **phase** of the particle undergoing the disturbance is a measure of its position in respect to its equilibrium position. Two or more waves from a disturbance may reinforce one another when they are changing in the same direction, or tend to cancel one another when varying in the opposite directions. When two or more waves have their displacement at all times in the same direction, they are said to be in phase with one another. See Fig. 11.1.

When interference occurs between waves, the two waves must remain exactly in phase for constructive interference (waves reinforce each other) and 180° out of phase for destructive interference (waves completely cancel each other). For complete destructive interference, the two waves must have the same amplitude. They must also have the same wavelength and be traveling in the same direction.

The above conditions can be obtained with the use of a single source of light. The waves spreading out from the source as shown in Fig. 11.2 fall upon a double slit S_1 and S_2. Thus, the waves spreading out from sources S_1 and S_2 have the same wavelength, the same amplitude, and are traveling in the same direction. The interference of these waves produces alternate bright and dark areas as shown in Fig. 11.2. The bright areas are regions where constructive interference is occurring and the dark areas are regions where destructive interference is occurring.

The relationship among the wavelength (λ), the distance (d) between S_1 and S_2 (called the slit separation), the path difference between two wave paths showing wave reinforcement (n), and the ratio of a value of (y) to a corresponding value of (x) (see Fig. 11.3) is given by the following equation:

$$n\lambda = d\left(\frac{y}{x}\right)$$

or

$$\lambda = \frac{d}{n}\left(\frac{y}{x}\right) \tag{11.1}$$

where $n = 1, 2, 3, 4, \ldots.$

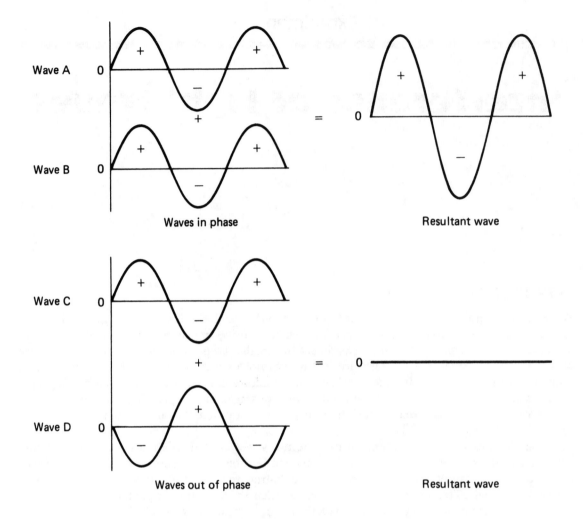

Figure 11.1 *Diagram illustrating the phase relationship between two waves that are in phase with one another, and two waves that are 180° out of phase. The resultant wave is also shown. Waves A and B are in phase. That is, their displacement at all times is in the same direction. Waves C and D are out of phase. That is, their displacement at all times is in the opposite direction. The waves shown have the same amplitude.*

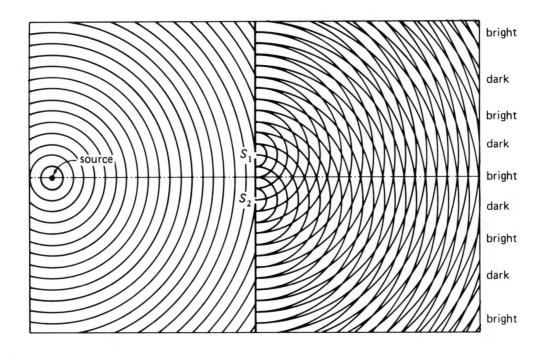

Figure 11.2   *Diagram illustrating the interference pattern of bright and dark regions produced by waves interfering with one another. Since the waves from S_1 and S_2 are from the same source, they have the same wavelength, the same amplitude, and are traveling in the same direction.*

LEARNING OBJECTIVES

After completing this experiment, you should be able to do the following:

1. Define and explain phase as it relates to wave motion.
2. Distinguish between constructive and destructive interference.
3. Calculate the wavelength of wave motion from data obtained in the experiment.

APPARATUS

Millimeter rule, two Keuffel and Esser wave-pattern transparencies, sheet of white paper, hand calculator.

85

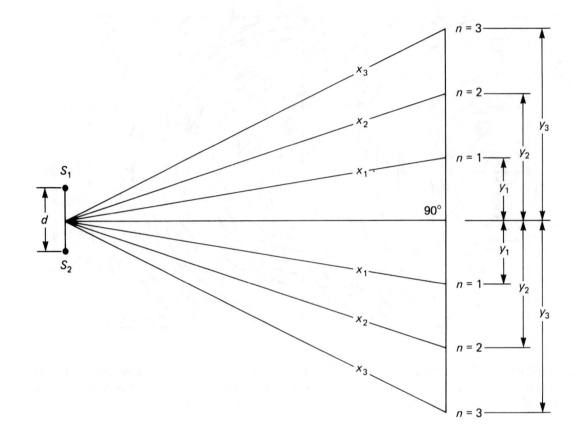

Figure 11.3

PROCEDURE 1

1. Take a sheet of white paper and place it on your laboratory table with the long side parallel to the edge of the table next to you. Draw a line 2 cm long near and parallel to the left edge of the paper.
2. Draw a dot at each end of the 2-cm line. These dots represent two sources of waves similar to S_1 and S_2 as shown in Fig. 11.2.
3. From the midpoint on the line between the two dots, draw a perpendicular line (call it the $n = 0$ line) to the right across the paper to a point near the edge of the paper. See Fig. 11.3.
4. Position the two plastic transparencies of concentric circles so that the center of the circles coincides with the two dots. This will produce an observable interference pattern.
5. Slightly rotate the two transparencies as a unit, if necessary, until the bright region near the perpendicular line ($n = 0$) is spaced evenly on each side of the perpendicular line.
6. With a pen, place a mark on the white paper at the right edge of the transparencies to coincide with the centers of the first, second, and third bright regions above and below the central bright region above the perpendicular line ($n = 0$). See Fig. 11.3.
7. Remove the transparencies and draw lines $x_1, x_2,$ and x_3 from the midpoint of the 2-cm line to the marks made in Step 6. Also draw lines $y_1, y_2,$ and y_3.
8. Label all x lines and all y lines. See Fig. 11.3.
9. Measure with the millimeter rule the length of all x and y lines, and record their values in Data Table 11.1.
10. Calculate the wavelength (λ) using Eq. 11.1. Record wavelength in Data Table 11.1.

PROCEDURE 2

Repeat Procedure 1 with the distance between S_1 and S_2 equal to 3 cm. Record the data in Data Table 11.1.

Data Table 11.1

d (cm)	n	y (cm)	x (cm)	λ calculated (cm)
2	1	$y_1 =$	$x_1 =$	
2	2	$y_2 =$	$x_2 =$	
2	3	$y_3 =$	$x_3 =$	
3	1	$y_1 =$	$x_1 =$	
3	2	$y_2 =$	$x_2 =$	
3	3	$y_3 =$	$x_3 =$	

Measured value for 20 wavelengths = _____ cm

1 wavelength = _____ cm

PROCEDURE 3

The wavelength of the concentric wave pattern on the transparencies can be measured directly with the millimeter rule. The accuracy of the measurement can be increased by measuring several wavelengths, rather than just one, then dividing by the number of wavelengths measured.

1. Measure the distance for 20 wavelengths and record in Data Table 11.1.

2. Determine the average value of the calculated wavelength.

3. Determine the percent difference between the average calculated wavelength and the measured wavelength.

QUESTIONS

1. Define and explain phase as it relates to wave motion.

2. When are two waves in phase at all times with one another?

3. When referring to wave motion, what is the meaning of the expression *out of phase*?

4. What three conditions must be met for complete destructive interference between two waves?

5. Distinguish between slit width and slit separation.

6. How does the wavelength vary with slit separation?

Mirrors, Lenses, and Prisms

INTRODUCTION

That region of the electromagnetic spectrum to which the eye is sensitive is known as light. At the red end of the spectrum, the wavelength of the visible radiation is 7.5×10^{-9} cm, and at the violet end, 4.3×10^{-5} cm. These numbers represent the approximate range of visible wavelengths. The boundaries between the different regions of the electromagnetic spectrum are not sharp; there is an overlapping. The spectrum range is sometimes given in frequencies instead of wavelengths. The relationship among frequency, wavelength, and velocity is given by the equation

$$v = \lambda f$$

where v = velocity of electromagnetic radiation,

λ = wavelength, in centimeters,

f = frequency, in cycles per second.

Electromagnetic waves confined to a vacuum or to a single medium will travel in a straight line. When waves traveling in one medium come to the boundary of a second medium, a change in direction takes place. If the electromagnetic radiation remains in the first medium, the change in direction is called **reflection**. If the radiation passes into and through the second medium, a change in velocity takes place. A change in direction, caused by the change in velocity, is called **refraction**. The laws relating to reflection and refraction are as follows: The first law of reflection states that when the reflection of light is from a plane specular surface, the incident ray, the reflected ray, and the normal to the surface at the point of contact are all in the same plane. The second law of reflection states that the angle of reflection is equal to the angle of incidence. The first law of refraction states that the incident ray, the refracted ray, and the normal to the surface are all in the same plane. The second law of refraction states that the sine of the angle of incidence divided by the sine of the angle of refraction is a constant. This second law is known as Snell's law.

LEARNING OBJECTIVES

After completing this experiment, you should be able to do the following:

1. Define reflection and refraction.
2. State the two laws of reflection.

3. State the two laws of refraction.
4. Determine experimentally the position, size, and characteristic features of images formed by concave mirrors and convex lenses.
5. Draw ray diagrams for concave mirrors and convex lenses.
6. Determine experimentally the color order of the visible part of the electromagnetic spectrum.

APPARATUS

Glass plate (4 in × 6 in), concave mirror, convex lens, meter stick, candle, screen (card), holders for mirrors and lenses.

PROCEDURE 1

Step 1 Place an 8 1/2 in × 11 in sheet of white paper on the laboratory table with the glass plate positioned in the center, as shown in Fig. 12.1. Position a small candle (of height less than 2 in) at the edge of the paper, as shown. With your eye in Position 1, as shown in the figure, you will observe an image of the candle in the plate glass. The plate glass is reflecting light and serving as a mirror. If you observe the top of the flame from a point slightly to the left of Position 1 (Position 2), you will see two images in the glass (again see Fig. 12.1). How far behind the glass is the image? One way to determine this is to hold a pencil behind the glass while viewing the image as shown in Fig. 12.1. Place the pencil so that the image of the flame appears to be in the body of the pencil. Measure the distance from the center of the pencil to the face of the glass, and record the distance in the space provided. Be sure to measure to the face (side) of the glass nearest to the image you located in the body of the pencil.

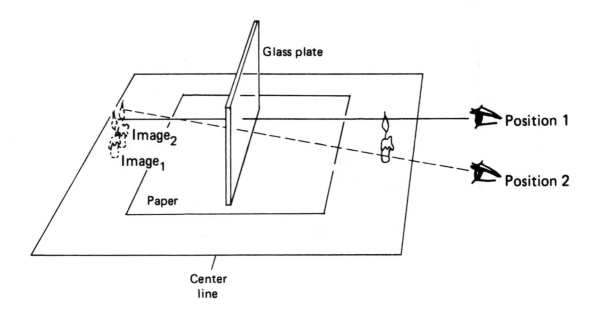

Figure 12.1

Distance of farthest image from the face of the glass nearest the image .... _____ cm

Measure the distance from the center of the candle to the face of the glass nearest the image.

Distance of candle from the glass .... _____ cm

How do the two distances compare? .... _____

Move the candle toward the mirror. What happens to the image? Measure and compare their distances from the face of the mirror.

Image distance .... _____ cm

Candle distance .... _____ cm

Comparison .... _____

Step 2 Mount the concave mirror and screen (card) on the meter stick, as shown in Figs. 12.2 and 12.3. Determine the focal length of the mirror as follows. Place the mirror at some fixed position on the meter stick; then move the screen toward or away from the mirror until a clear image of a distant object outside the window appears on the screen. The distance between the mirror and the screen, when adjusted for a clear image, is the focal length. Take three trials and calculate the average. Record in Data Table 12.1.

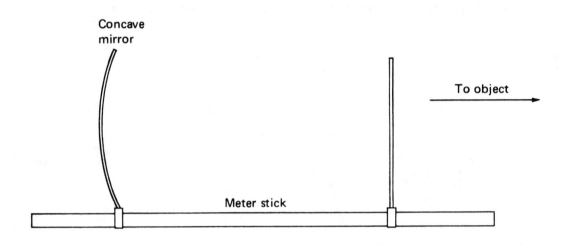

Figure 12.2 *Side view of optical bench.*

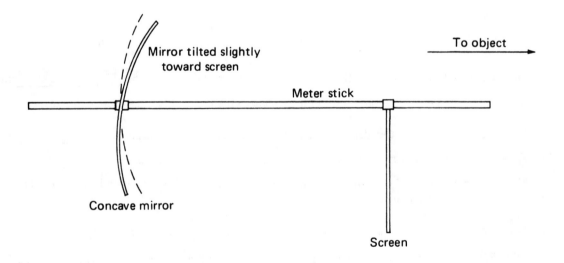

Figure 12.3 *Top view of optical bench.*

Data Table 12.1

Trial	Focal length of mirror
1	
2	
3	
Average	

Step 3 Images formed by mirrors and lenses are real or virtual, smaller or larger than the object, and located in front of it or behind the optical device. A **real image** is one that can be focused on a screen. A **virtual image** cannot be focused on a screen; to see it one must look into the mirror or lens. Mount the concave mirror, screen, and candle on the meter stick, as shown in Fig. 12.4. The diagram shows the relationship between the focal length f and the radius of curvature R of the mirror. The radius of curvature equals two times the focal length. In Step 2, you determined the position of the image when the object is at *plus* infinity. Use the data from Step 2 to answer Line 1 of Data Table 12.2. Starting with the candle at some distance greater than $2R$, determine the position of the image as the candle is moved toward the face of the mirror. When entering the data in the table provided, record the position of the image as being at f, at R, between f and R, behind the mirror, etc. Record the image position in a manner similar to the way the candle's position is given.

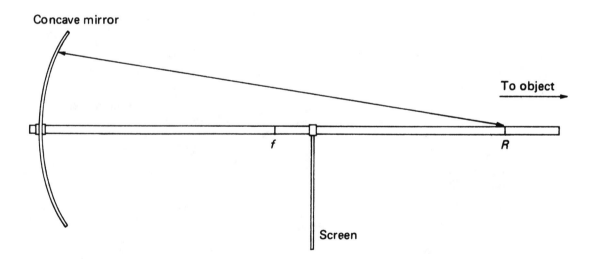

Figure 12.4 *Top view of optical bench.*

Step 4 The focal length of a lens is determined by the radius of curvature and the index of refraction of the glass. Mount the double convex lens as shown in Fig. 12.5. Determine the focal length of the lens, using some object outside the window and adjusting the screen for a clear image. Consider this object at plus infinity and record the data called for in Line 1 of Data Table 12.3. Place the meter stick on support holders on the laboratory table. Position the candle at the distance in front of the lens called for in the data table. Record the data; then move the candle toward the lens, obtaining information to complete the data table.

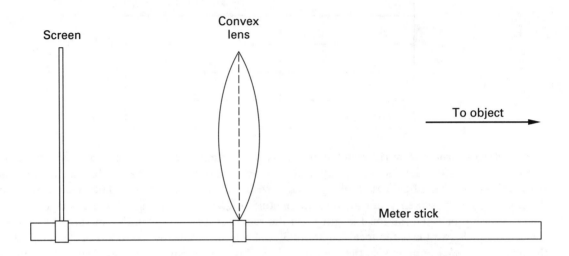

Figure 12.5   *Side view of optical bench.*

Step 5 Position the prism with respect to the light source, as shown in Fig. 12.6, and record the colors observed in the spectrum.

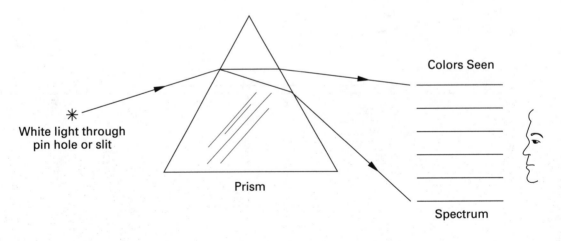

Figure 12.6

Data Table 12.2 Data Table for Mirror

Position of candle	Information on image			
	Position	Real or virtual	Erect or inverted	Larger or smaller
+ infinity (at least 8f)				
At 4f				
At 2$f = R$				
Between R and f				
At f				
Between f and mirror				

Data Table 12.3 Data Table for Lens

Position of candle	Information on image			
	Position	Real or virtual	Erect or inverted	Larger or smaller
+ infinity (at least 8f)				
At 4f				
At 2f				
Between 2f and f				
At f				
Between f and lens				

PROCEDURE 2

Use the following procedure to draw a diagram to locate the image formed by a concave mirror.

1. Draw a ray from the object to the face of the mirror parallel to the optical or principal axis of the mirror. When reflected by the mirror, this ray will pass through the focal point. See Fig. 12.7 below.

2. Draw a second ray from the same point on the object to the face of the mirror so that the ray passes through the radius of curvature. This ray will be reflected back on the same path as the entering ray since the entering ray came in on the normal. The angle of incidence is zero, therefore the reflected angle is zero.

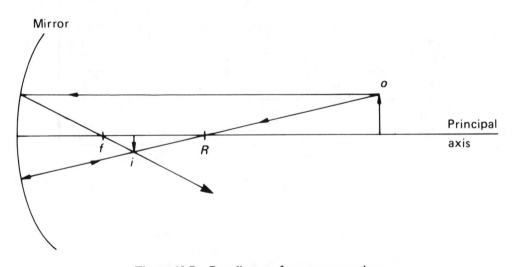

Figure 12.7 *Ray diagram for a concave mirror.*

PROCEDURE 3

Use the following procedure to draw a diagram to locate the image formed by a convex lens.

1. Draw a ray from the object to the lens parallel to the optical or principal axis of the lens. Then extend the line through the lens to and through the focal point on the other side of the lens. See Fig. 12.8 below.

2. Draw a second ray from the same point on the object to the lens such that the ray passes through the optical center of the lens. This ray will not be deviated from its original path. In other words, draw a straight line from the point on the object through the center of the lens to the other side of the lens. The image is located at the intersection of the two rays.

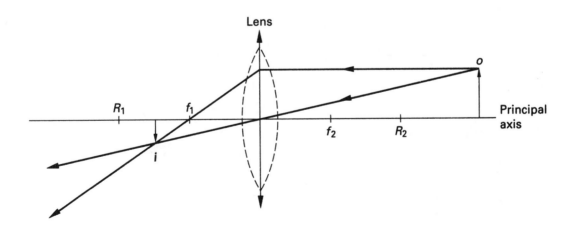

Figure 12.8   *Ray diagram for a convex lens.*

Refer to your textbook or ask the instructor about drawing ray diagrams when the object is between the focal point and the mirror or lens.

EXERCISES

1. Draw ray diagrams for the concave mirror when the object is (1) at 4*f*, (2) between *R* and *f*, (3) between *f* and mirror. Refer to Procedure 2.

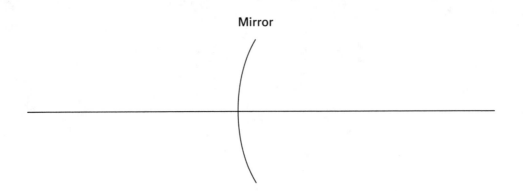

Mirror

2. Draw ray diagrams for the convex lens when the object is between (1) 2*f* and *f*, (2) *f* and lens. Refer to Procedure 3.

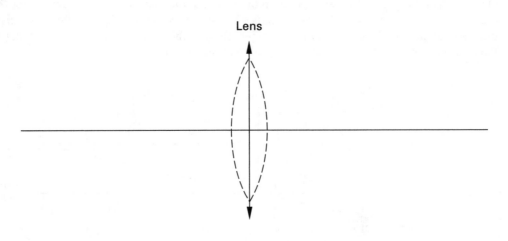

Lens

QUESTIONS

1. Distinguish between reflection and refraction.

2. State the two laws of reflection.

3. State the two laws of refraction.

4. Why were two images observed in the glass plate used in Procedure 1 Step 1?

5. Distinguish between a real and a virtual image. Give an example of each.

6. At what distance must the object be placed in front of a concave mirror for the image to appear the same size as the object? Draw a ray diagram to illustrate your answer.

7. At what distance must the object be placed on the principal axis of a double convex lens for the image to appear the same size as the object? Draw a ray diagram to illustrate your answer.

The Refracting Telescope

INTRODUCTION

Lenses made of glass refract light. That is, the lens changes the direction of light rays because the velocity of light changes when it passes through a different medium. Parallel rays of light entering one face of a glass lens will be refracted and converge to a point (called the focus or focal point) on the back side of the lens. See Fig. 13.1. When the entering rays are parallel to the principal or optical axis of the lens (Fig. 13.1a), then the distance from the center of the lens to the focal point is known as the **focal length** of the lens. Parallel rays of light that are coming in at an angle to the optical axis, entering one face of the lens, will come to a point or focus on the back side of the lens at the focal plane. See Fig. 13.1b and Fig. 13.1c. Thus, if a card or screen were placed on the back side of the lens at the focal point, an image of some distant object located in front of the lens would be seen on the screen. However, if you put your eye at the focal point, you would not see the image of the object. For example, if the object were the moon, your eye would only see a bright lens. To see an image, the converging rays of light must be changed to parallel rays so that your eye's lens will be able to focus them on the retina. This can be accomplished by using an additional lens.

Fig. 13.2 illustrates a simple refracting telescope. Light from some distant object enters the telescope through lens L_1, called the **objective lens**, forming an inverted image at the focal plane. The diameter of this lens is usually referred to as the **aperture**. The greater the aperture the greater the light-gathering power of the telescope. The objective lens, which has a long focal length (f_o), is fixed in position. Lens L_2 is called the **eyepiece**. This lens is set in a movable tube so that the distance between it and the objective lens can be varied. The eyepiece has a small diameter and a short focal length (f_e) in comparison to the objective. The image formed by the objective becomes the object for the eyepiece. The eyepiece forms a clear, distinct, magnified, inverted, and virtual image of the distance object. See Fig. 13.3.

The magnification of the simple two-lens telescope is a function of the focal length of the two lenses. The relationship can be written as

$$\text{magnification} = \frac{\text{focal length of the objective}}{\text{focal length of the eyepiece}} = \frac{f_o}{f_e}$$

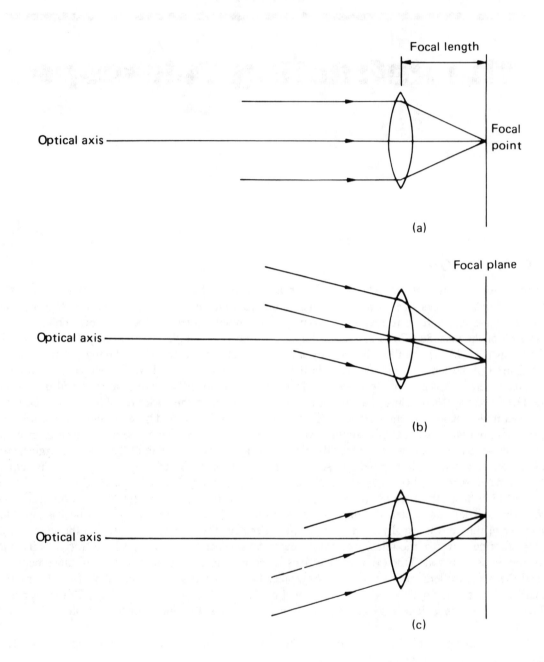

Figure 13.1 *Diagram illustrating the converging of parallel rays of light by a convex lens (a) when the incoming parallel beam is parallel to the optical axis of the lens, (b) and (c) when the incoming parallel beam enters the lens at an angle to the optical axis.*

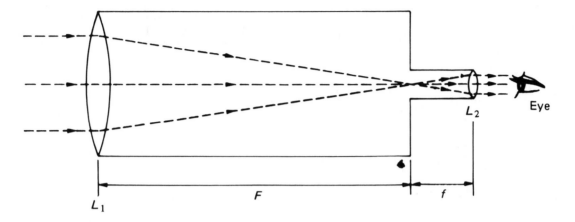

Figure 13.2 *Diagram illustrating a simple two-lens refracting telescope.*

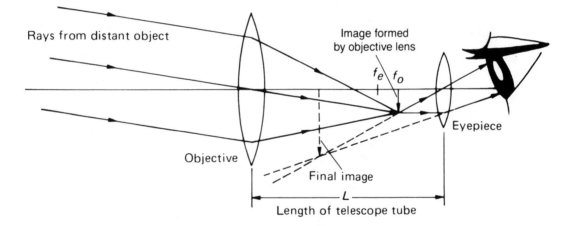

Figure 13.3 *Ray diagram showing final image of some distant object.*

LEARNING OBJECTIVES

After completing this experiment, you should be able to do the following:

1. Define the terms focal length, objective lens, eyepiece, and aperture.
2. Determine theoretically and experimentally the magnification of a simple two-lens telescope.
3. Calculate the light-gathering power of a telescope.
4. Calculate the resolving power of a telescope.

APPARATUS

Meter stick, objective lens and holder, eyepiece and holder, screen and holder (a small index card), a colored filter and holder. The filter is not necessary, but is helpful, when determining the magnification of the telescope.

PROCEDURE

1. Mount the objective lens (the large-diameter lens) in its holder and place it on the meter stick at the 10-cm mark. Mount the cardboard screen in its holder and place it on the meter stick near the 30-cm mark. Go to an outside window and point the meter stick toward the outside with the lens toward the window and the screen next to you. Adjust the screen back and forth until you obtain a clear image on the screen of some distant object (tree or building) outside the laboratory. If the experiment is being done at night, choose a street lamp or a neon sign as the object. Once a clear image is obtained, measure the distance from the center of the lens to the screen. This is the focal length of the lens. Make three independent trials; record them in Data Table 13.1 and calculate their average.

2. In the same way, obtain the focal length of the eyepiece lens. Again make three independent trials; record them and calculate their average.

3. Mount the two lenses in their holders on the meter stick. Remove the screen and its holder from the stick entirely. Place the lenses a distance apart nearly equal to the sum of their focal lengths, and look through the eyepiece lens and the objective lens at some distant object outside the laboratory window. Move the eyepiece lens slightly until the object is clear and sharp. Now measure and record the distance between lenses. Make two additional trials and record the data. Calculate the average of the three trials.

4. Two students are needed to do this part of the experiment. One student will use the telescope at the end of the laboratory opposite the blackboard. The other student will be at the blackboard. Draw a 20-cm vertical line on the blackboard with white chalk. Next draw a horizontal line 5 cm long about 5 cm to the right of the top of the vertical 20-cm line to indicate where the top of the 20-cm line is located. Set the lenses of your telescope the average distance apart found in (3) above and mount the green filter between them. Look at the 20-cm vertical line through the telescope with one eye closed. You will see a magnified vertical green line if the filter is used. Otherwise, you will see a magnified white line. Continue looking through the telescope with one eye closed. Now open the other eye. Both eyes are now open. One eye is seeing a magnified green image and the other eye is seeing the white 20-cm vertical line as drawn on the blackboard. This is not easy for some to do. Keep trying. Next adjust the elevation of the telescope so that the top of the green magnified line is at the same level as the unmagnified vertical white line. With this done instruct your partner at the blackboard to make a horizontal line at the bottom of the green magnified image. The partner may have to draw and erase a few lines before you accomplish this. Once this is done, measure the distance in centimeters between the two horizontal lines and record in the data table. Divide the value by 20. This is your experimental value of the magnification of the telescope. Record the value in the data table.

QUESTIONS

1. What is the simplest way to increase the magnifying power of the two-lens telescope?

107

2. Why was an object outside the window of the laboratory used when the focal lengths of the two lenses were determined?

3. The brightness of the image of an object increases in direct proportion to the area of the objective lens or the square of its diameter. The objective lens collects light from a large beam of light, then concentrates it into a small beam for the eye to see. The light-gathering power of the telescope is defined as the ratio of the light per unit time entering the eye by way of the telescope to the light from the object that enters the unaided eye per unit time. The relationship can be written as

$$\text{light-gathering power} = \frac{(\text{aperture of the telescope})^2}{(\text{aperture of the eye})^2}$$

The aperture of the human eye is 1/4 in.

$$\therefore \text{light-gathering power} = 16 \, (\text{aperture of the objective})^2$$

(a) Calculate the light-gathering power of the telescope constructed in the laboratory. Show your work. (*Note:* aperture size must be in inches.)

(b) Calculate the light-gathering power of the 200-in telescope on Mount Wilson. Show your work.

4. In addition to its light-gathering power a telescope is rated on its resolving power or its ability to separate two objects that are close together. A relationship derived from diffraction theory can be written as

$$\text{resolving power} = \frac{5 \text{ inches}}{\text{aperture of the objective in inches}}$$

(a) Calculate the resolving power of the telescope constructed in the laboratory. Show your work.

(b) Calculate the resolving power of the 200-in telescope on Mount Wilson. Show your work.

Data Table 13.1

1.	Position of objective lens on meter stick	_____	cm	_____	cm	_____	cm
	Position of screen on meter stick	_____	cm	_____	cm	_____	cm
	Focal length of objective	_____	cm	_____	cm	_____	cm
	Average focal length of objective	_____	cm				
2.	Position of eyepiece lens on meter stick	_____	cm	_____	cm	_____	cm
	Position of screen on meter stick	_____	cm	_____	cm	_____	cm
	Focal length of eyepiece	_____	cm	_____	cm	_____	cm
	Average focal length of eyepiece	_____	cm				
3.	Position of objective on meter stick	_____	cm	_____	cm	_____	cm
	Position of eyepiece on meter stick	_____	cm	_____	cm	_____	cm
	Distance between lenses	_____	cm	_____	cm	_____	cm
	Average distance between lenses	_____	cm				
4.	(a) Vertical height of image _____ cm						
	(b) Vertical height divided by 20 = _____						

CALCULATIONS

1. Sum of average focal lengths of lenses   _____

    Distance between lenses (from 3 in Data Table 13.1)   _____

    Percentage difference between above two values   _____

2. (a) Magnification [from 4(b)]   _____

$$\text{(b) Magnification} = \frac{\text{focal length of objective}}{\text{focal length of eyepiece}} = \qquad _____$$

(c) Percentage error [use 2(b) as correct value]   _____

Color

INTRODUCTION

Color is a perception made possible by the complex structure of the human eye. The eye, as you know, is sensitive to light, which allows us to see objects. We detect light either directly from glowing objects, or reflected from the surface of an object that does not itself glow. The color that we see is dependent on the wavelengths of the light that are produced by the glowing object or that are reflected from the nonglowing object. White is the presence of all colors (i.e., all wavelengths of light) in the same proportional amount as in the light produced by our nearest star, the Sun. Black is the absence of all light; if an object appears black, it does not produce or reflect any visible light.

In this experiment you will look at several aspects of color by doing six short subexperiments ranging from simple observation to the mixing of pigments. Because each stands alone, the subexperiments can be done in any order. In most cases you will move from one experimental setup to another, performing one subexperiment at each location.

LEARNING OBJECTIVES

After completing this experiment, you should be able to do the following:

1. Describe the spectrum of white light after it has passed through a diffraction grating.
2. List the three primary colors of light and tell what colors are produced by the various combinations of these colors.
3. Explain what occurs when plastic color filters of various colors are combined.
4. Show how to best match color paint-chips to cloth swatches of various colors.
5. Mix paint pigments to form the secondary colors in the artist's subtractive color theory, and learn about tints, tones, and shades.
6. Observe and explain the effects of colored lighting on a variety of colors.

APPARATUS

White-light source and diffraction grating, Singerman light box or Fisher black-ray box with color filters for the primary colors of light, set of 12 to 20 plastic color filters, several paint-chips from a paint store in colors that approximate 4 to 6 pieces of colored cloth, 4 to 6 pieces of colored cloth, colored tempera paints (liquid, in red, blue, green, yellow, black, and white), wooden mixing sticks and brushes, plastic mixing sheets, newspapers to paint over, 6 to 8 small cups or beakers for water, one high-intensity lamp with red, blue, and green plastic transmission filters.

PROCEDURE 1

Look at the white-light spectrum produced by a diffraction grating. List below the hues that you can see in this spectrum.

PROCEDURE 2

Turn on each of the three light sources one at a time and record the color produced by each one, as seen on a white surface.

List the three *primary* colors of light.

1. _____

2. _____

3. _____

Now turn on two of the light sources at a time and record the color observed on the screen. List the *secondary* colors of light that you see.

red + blue =   _____

green + blue = _____

red + green = _____

Turn on all three light sources at once and record what you observe on the screen.

red + green + blue = _____

Carefully take the green filter out and turn on each of the other light sources, one at a time, with the white light. Record the results below.

red + white = _____

blue + white = _____

PROCEDURE 3

You will find several plastic colored filters on the table. Combine these in pairs on top of a white sheet of paper. For six pairs, record the color you observe in the region where the two color filters overlap. Record the results below.

Trial	Color 1	Color 2	Combination color
1			
2			
3			
4			
5			
6			

PROCEDURE 4

Keeping the cloth and color chart well-separated, take a sample of cloth and look at its color. Now look at the paint chip provided and pick out the color that you think best matches the cloth. Record its designation number below.

Now place the cloth sample and the paint-chip side by side and see if you still agree with your original choice. Record your second choice below.

Why do you think there might be a difference between your choices?

Do the same for the other cloth samples provided. Record the closest color-code number:

1.   2.   3.

Did everyone in your group agree on the choice of color-code number for the paint-chip that they thought best matched the cloth? Can you explain why not? (If your group is typical, you probably did not all agree.)

PROCEDURE 5

On the Color Work Sheet provided, fill in circles with the appropriate color or combination of colors. Place one or more *small* dabs of paint on a plastic mixing sheet with the wooden sticks provided and then mix it with a brush to form a uniform color. This color can then be applied to the appropriate circle. (*Note:* The circles do *not* need to be completely filled in.)

When making blue tints, start with pure blue and add a little more white to each succeeding dot (Row 3).

When making red shades, start with pure red and add a little more black to each succeeding dot (Row 4).

Fill in the circles on Row 5 with a series of grey tones. To make grey tones, start with pure white and add a little more black to each succeeding dot.

Mix up a medium grey tone and paint a dot in the center of a piece of white paper and another dot using the same paint mixture in the center of a piece of black paper.

Question: Do there appear to be any differences in color between the two grey paint dots? Explain.

PROCEDURE 6

1. Place a red colored filter over a high-intensity lamp and allow the light to fall on the piece of white paper on which you painted the dots in Procedure 5. Note any major changes in the colors of the dots.

2. Why do you think the changes in observed color occur when the same light is shining on the different shades of grey?

3. Now place a blue filter over the light and record any major changes in the color of the dots.

4. Repeat the procedure with a green filter and record the color changes of the dots.

COLOR WORK SHEET

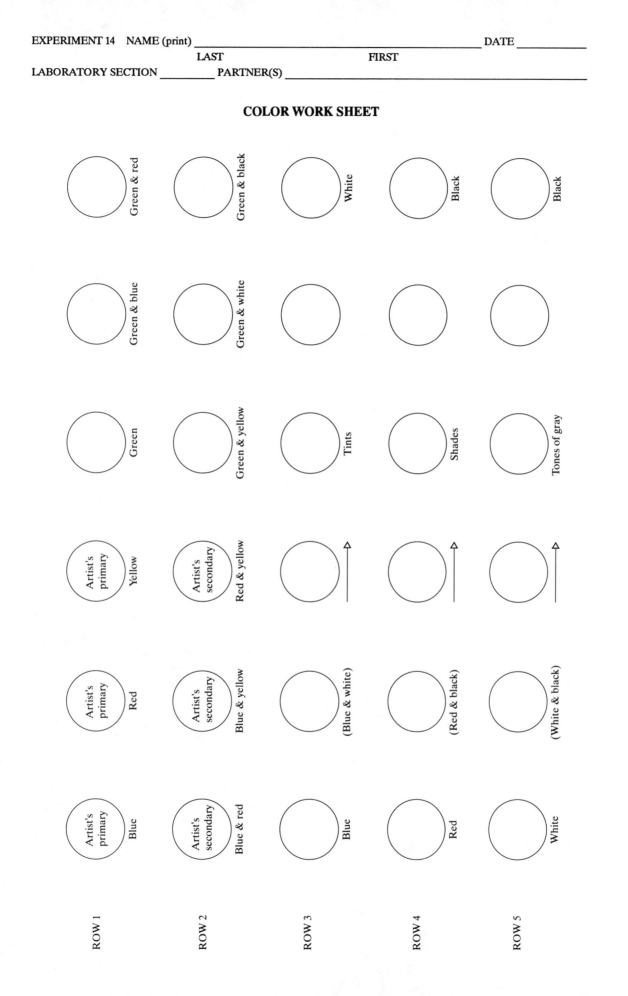

Static Electricity

INTRODUCTION

According to modern theory, all matter consists of very small particles called atoms that are composed in part of negatively charged particles called electrons, positively charged particles called protons, and particles called neutrons that carry no electric charge.

Electrons and protons possess the four fundamental properties of matter—mass, length, time, and electric charge. **Charge** refers to the force field that the electron and the proton possess. To differentiate the force field from a gravitational or nuclear force field, we call it an **electric** field.

The magnitudes of the electric force field on an electron and a proton are equal, but their polarity or sign is different. When the same number of electrons and protons are present on a body, the total net charge is zero. That is, there is no apparent force field present. We have a neutral situation. Hence, we call the small particle with zero electric charge a **neutron.**

Electric charge is measured in units called coulombs and designated by the symbol (q). Negative charges are designated by a (–) sign and positive charges by a (+) sign. All electrons have the same charge (-1.6×10^{-19} coulomb), and all protons have the same charge (1.6×10^{-19} coulomb). Presently all particles found in nature have multiples of this amount of electric charge. Some scientists, however, believe in the existence of particles called **quarks**. Quarks have charges of $-2/3$, $-1/3$, etc. In the currently accepted theory, quarks combine (three at a time) to form protons and neutrons. So far, however, not a single free quark has been found.

The relationship between charges is given by Coulomb's law. The law states:

The force of attraction between two unlike charges or the force of repulsion between two like charges is directly proportional to the product of the two charges and inversely proportional to the square of the distance between them.

The general law of signs for electric charges states that unlike charges attract and like charges repel. See Fig. 15.1.

Two unlike charges attract

Two like charges repel

Figure 15.1 *Diagram illustrating the law of signs for electrical charges.*

LEARNING OBJECTIVES

After completing this experiment, you should be able to do the following:

1. State the four fundamental properties of the electron and proton.
2. State Coulomb's law for electric charges.
3. State the general law of signs for electric charges.
4. Identify certain properties of static electricity.

APPARATUS

Aluminum or gold leaf electroscope, two hard-rubber or ebonite rods, a wool cloth or fur, two glass rods, support stand with holder, silk cloth, toy balloons.

PROCEDURE 1

Place an electrostatic charge on one of the rubber rods by rubbing it with wool or fur. Suspend the rod as shown in Fig. 15.2. Place a charge on the other rubber rod and bring it near one end of the suspended charged rod. Record what you observe as answer A in Data Table 15.1. Bring the rubber rod near the other end of the suspended rubber rod. Record what you observe as answer B in Data Table 15.1. Bring the fur or wool near the suspended rubber rod. Record what you observe as answer C in the data table. Place an electrostatic charge on one of the glass rods by rubbing it with the silk cloth. Bring the glass rod near one end of the suspended rubber rod. Record what you observe as answer D in the data table. Bring the glass rod near the other end of the suspended rubber rod. Record what you observe as answer E in the data table. Bring the silk cloth near the suspended rod. Record what you observe as answer F in Data Table 15.1. What can you conclude from the observations in the above procedure? Record your answer as answer G in Data Table 15.1.

PROCEDURE 2

The electroscope is an electrical device used to detect and determine the polarity (sign) of charged bodies. In its simplest form, it consists of two gold or aluminum leaves held by a conductor (usually copper or brass) and housed in a container that protects the leaves from air currents.

Figure 15.2   *Suspended rubber rod free to move in a horizontal plane.*

121

Data Table 15.1

Answer A	
Answer B	
Answer C	
Answer D	
Answer E	
Answer F	
Answer G	

Discharge the electroscope by touching the brass knob with your finger. A path is thereby provided for any charge on the leaves to discharge into your body. Only the negative charged particle (the electron) is free to move. The positive charged proton is part of the nucleus of an atom, and the proton is not free to move. Thus, when charges are added to or taken from a body, it is electrons that are added or taken away. Any charge on the glass or rubber rods can be removed by rubbing your hand completely over the surface; the rods are insulators, and therefore contact with all the surface is necessary to remove all charges. With no charge on the rubber rod, there should be no effect on the leaves of the electroscope when the rubber rod is brought into the vicinity of the electroscope. See Fig. 15.3a.

A charged electroscope is needed to detect and determine the polarity (sign) of charged objects. One of two methods can be used to charge an electroscope. One method is by contact, or simply touching the aluminum or brass knob of the electroscope with an object carrying a charge. This method is not used in this experiment because it is difficult to control the amount of charge transferred to the leaves of the electroscope. **Caution:** Do not touch the brass knob with any charged object. The second method is by induction. To charge the electroscope by induction, proceed as follows: Stroke the rubber rod with wool or fur, thereby placing a negative charge (excess electron) on the rubber rod. Electrons are removed from the wool or fur and transferred to the rubber rod. As you bring the charged rod into the vicinity of the electroscope, you should observe the electroscope leaves diverging. See Fig. 15.3b. While holding the rod in a position that causes the leaves to spread apart, but not touching the rod to the knob, touch the knob of the electroscope with a finger of your free hand. See Fig. 15.3c. Electrons on the leaves will flow into your body. Remove your finger from the knob, then remove the rubber rod from the vicinity of the electroscope. Because electrons on the leaves have been removed, the electroscope will be positively charged. See Fig. 15.3d. As the rubber rod is brought near or taken away from the charged electroscope, you will observe that the leaves of the electroscope spread farther apart or come closer together. As the negative rubber rod is brought close to the knob of the electroscope, electrons on the knob are repelled into the leaves and neutralize part or all of the positive charges. Thus, the leaves come closer to one another or collapse completely, depending on how close the rubber rod is brought near the knob of the electroscope. When the rod is taken away, the charge distribution returns to the electroscope as originally charged by induction.

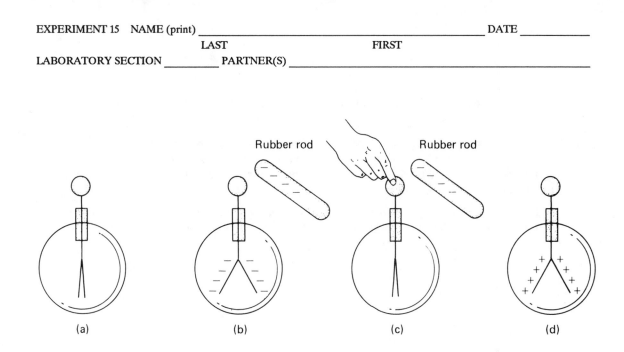

Figure 15.3 *Charging an electroscope by induction. See the text for an explanation.*

PROCEDURE 3

Using the charged electroscope, complete the following:

1. Determine the charge on a glass rod after rubbing it with silk. The charge on the glass rod is _____ (positive or negative). The charge on the silk cloth is _____.

2. Rub each of two inflated toy balloons with wool or fur. Will they attract each other? Explain why.

3. Determine the charge on each balloon.

4. Determine the charge on the wool or fur.

123

5. What kind of charge will a sheet of paper have after it has been rubbed with wool or fur? _____ with silk ? _____

6. Obtain a few (four or five) pieces of paper 1 or 2 cm in size. Remove any electric charge by touching them with your hands. Test for any charge on the pieces of paper by bringing them near the knob of the charged electroscope. The pieces of paper are neutral. That is, they have zero charge. Place them on the table free of any electric charge.

(a) Place a negative charge on a rubber rod by rubbing it with wool or fur. Bring the charged rubber rod near the zero-charged pieces of paper. What do you observe? Explain.

(b) Repeat using a glass rod and a silk cloth. What do you observe? Explain.

QUESTIONS

1. What kind of a charge can be placed on an electroscope by induction using a glass rod rubbed with silk? Explain.

2. How many different kinds of charges did you observe the effects of in this experiment?

3. State the law of signs for electric charges.

4. Two metal spheres mounted on insulated stands are placed in contact with one another as shown
   in Fig. A below. A rubber rod carrying a negative charge is placed near the left side of sphere A.
   The spheres are then separated while the rubber rod is held near sphere A. See Fig. B below. Will
   the spheres possess an electric charge? If yes, what kind of charge will be on each? Explain why
   each is charged with the kind of charge you have indicated.

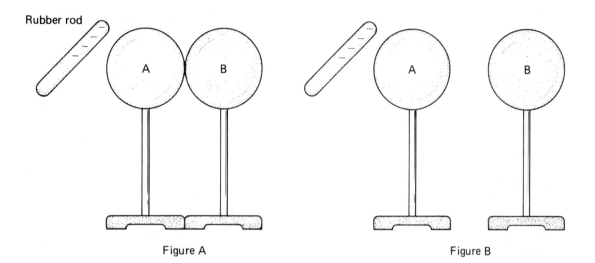

Figure A   Figure B

125

Magnetism and Electromagnetism

INTRODUCTION

The history of magnetism began with ancient civilization in Asia Minor in a region known as Magnesia, where rocks were found that would attract each other. These rocks were called **lodestones** or **magnets**.

The origin of the compass is unknown, but it is fairly certain that the first magnetic compass was the lodestone or leading stone used in the latter part of the thirteenth century.

Magnetism is the phenomenon of matter (matter that does not display an electric force field) to attract or repel other matter. This phenomenon is produced by a magnetic force field.

A magnetic force field is generated when electric-charged matter such as an electron, proton, or ion is put in motion. Thus, spinning electrons or electrons moving through a conductor generate a magnetic field of force. No magnetic field exists around a stationary charged particle. That is, if the particle is not spinning or moving in some direction, no magnetic field is present. The magnetic force field possessed by the bar magnets used in this experiment is due to the motion of electrons in the atoms of iron or other elements used to make the magnet.

LEARNING OBJECTIVES

After completing this experiment, you should be able to do the following:

1. State the general law for magnetic poles.
2. State Coulomb's law for magnetic poles.
3. State the method for determining the direction of a magnetic field at a given position.
4. State the rule for determining the direction of a magnetic field generated by a moving charge.
5. Draw on a sheet of paper the magnetic lines of force produced by a bar magnet.
6. Determine the angle of dip for Earth's magnetic field.

APPARATUS

Two bar magnets, meter stick, ring stand or a vertical support rod for the laboratory table, pendulum clamp (Cenco number 72296 or similar support clamp), thread, dip needle, large sheet of paper, magnetic compass (small size), dry cell (1.5 V), a short piece (25 cm) of number 14 copper wire.

PROCEDURE 1

Suspend one of the bar magnets as shown in Fig. 16.1 so that it is free to move in a horizontal plane about a vertical axis. After removing any magnets or other items that might influence the movement of the suspended magnet, allow it to swing freely and come to rest. At rest, the suspended magnet has aligned its long axis with the Earth's magnetic field. That is, the magnetic field of the suspended magnet and the magnetic field of the Earth act on each other, and being free to move, the magnet orients its long axis parallel to the Earth's magnetic field. The end of the magnet pointing in the direction of geographic north is called the north-seeking pole of the magnet, or simply the north pole. The opposite end of the magnet is called the south pole.

If the suspended magnet fails to orient in a north-south direction with the end marked (N) pointing north, ask the instructor for assistance. The word north should be posted on one wall of the laboratory to inform you of the direction of geographic north.

The suspended magnet, which acts as a compass, may not point directly toward geographic north. In fact, in most places on the Earth's surface it will not point directly toward geographic north. The geographic north pole and the magnetic north pole are not located at the same place, but are several miles from each other. The angle that a compass needle deviates from geographic north is called the angle of declination.

Holding the second bar magnet in your hand, bring its north pole near the north pole of the suspended magnet. Record what you observe in Data Table 16.1 as answer A. Next, bring the north pole of the magnet in your hand near the south pole of the suspended magnet. Record what you observe in Data Table 16.1 as answer B. Bring the south pole of the magnet held in your hand first to the north pole of the suspended magnet, then to the south pole. Record what you observe in Data Table 16.1 as answers C and D, respectively. What can you conclude from the four observations concerning the attraction and repulsion of magnetic poles? Record your answer in Data Table 16.1 as answer E.

The general law for magnetic poles states that two unlike magnetic poles attract each other, and two like magnetic poles repel each other. How does this compare with your answer E?

Coulomb's law for magnetic poles states that two unlike magnetic poles attract each other and two like magnetic poles repel each other with a force directly proportional to the product of their pole strength and inversely proportional to the square of the distance between the poles.

Figure 16.1   *Suspended bar magnet free to move in a horizontal plane.*

Data Table 16.1 Procedures 1 and 2

Answer A	
Answer B	
Answer C	
Answer D	
Answer E	
Answer F	
Answer G dip angle =	

Data Table 16.1 Procedure 3

Position A	
Position B	
Position C	
Position D	
Position E	
Position F	

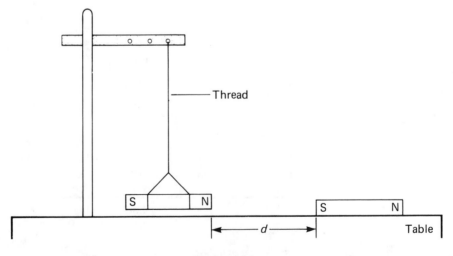

Figure 16.2

Loosen the pendulum clamp and lower the suspended bar magnet until it is just above the surface of the laboratory table, as shown in Fig. 16.2. Allow the suspended magnet to come to rest in the Earth's magnetic field. Make sure the second bar magnet does not influence the suspended magnet.

After the suspended magnet has stopped moving, place the second bar magnet 0.5 m north of the suspended magnet, as shown in Fig. 16.2. Displace the suspended magnet slightly from its equilibrium position, then notice how long it takes the magnet to come to rest. Move the bar magnet on the table so that the distance (d) is 0.25 m. Displace the suspended magnet slightly and notice how long it takes the magnet to come to its equilibrium position. Repeat with the distance (d) equal to 0.1 m.

What general conclusion can you make about the force that influences the suspended magnet in relationship to the distance between its magnetic pole and the magnetic pole of the magnet on the table? Record your answer in Data Table 16.1 as answer F.

PROCEDURE 2

A portion of the magnetic field surrounding a bar magnet can be plotted using a large sheet of paper to record the so-called lines of force, and a very small compass can be used to show the location and direction of the lines of force. See Fig. 16.3.

A magnetic force field is a vector quantity, and the direction of the force field at any given location, by definition, is the direction a unit north pole would tend to move when placed in the force field at that location. Since like poles repel and unlike poles attract, the direction of the force field is away from the north pole of the magnet producing the field and toward the south pole. See Fig. 16.3.

Position a bar magnet in the center of a large sheet of paper with its long axis in an east-west direction. (One sheet of paper is to be used by each group of students at a laboratory table.) Draw an outline of the magnet with a pen or pencil. Label the outline N and S, respectively. See Fig. 16.3. With the magnet in position, make a small dot near one end of the magnet; then place the small compass so that one end of the compass needle coincides with the dot. Make sure the compass needle rotates freely. Tap the top of the compass slightly to free a needle that tends to stick. Place a dot at the opposite end of the compass needle after the needle comes to rest. Move the compass until the end of the needle that originally coincided with the first dot coincides with the second. Continue this process until the compass is back to some part of the magnet or reaches the edge of the paper. See Fig. 16.3. Connect the points with a smooth line and draw arrows on each line to indicate the direction in which the north pole of the compass pointed.

Continue this process for a number of other lines until a symmetrical field is obtained. A diagram similar to that shown in Fig. 16.4 should be obtained. Notice on the diagram that no two lines cross. The points indicated by (O) are points where, due to the bar magnet, the magnetic field is equal and opposite to the horizontal component of the Earth's magnetic field. When a small compass is placed at either of these points, the needle of the compass will point in any direction.

The plot shown on the large sheet of paper illustrates the reaction between the magnetic field of the bar magnet and the horizontal component of the Earth's magnetic field.

The true direction of the Earth's magnetic field can be obtained with a dip needle. Ask the instructor for a dip needle and proceed as follows:

1. Remove all magnets and magnetic material that will influence the dip needle.
2. Adjust the dip needle so that it will rotate in the horizontal plane.
3. Allow the needle to reach equilibrium in a north-south line.
4. Adjust the dip needle so that it will rotate in the vertical plane.
5. Allow the needle to reach equilibrium.
6. Read the angle of dip. This is the angle the Earth's magnetic field makes with the Earth's surface.
7. Record the dip angle in Data Table 16.1 as answer G.

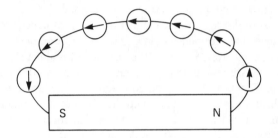

Figure 16.3 *Compass used to plot a magnetic field line of a bar magnet.*

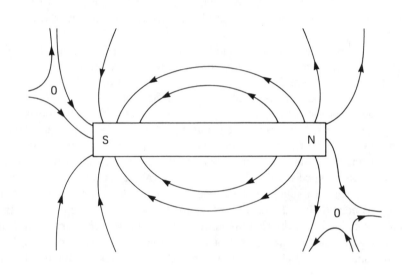

Figure 16.4

PROCEDURE 3

It was stated in the introduction that a magnetic force field is generated by a moving electric charge. This can be shown experimentally as follows:

1. Bend the number 14 wire to form a square loop as shown in Fig. 16.5.
2. Connect one end of the wire to the positive (+) terminal of the dry cell, but do not connect the other end of the wire to the negative (−) terminal.
3. Determine the direction of the magnetic force field around the wire by observing the compass needle when the compass is placed at the positions A through F as shown in Fig. 16.5. This is done by holding the compass in the labeled position, then momentarily touching the end of the wire to the negative (−) terminal of the dry cell. The time of contact to the negative terminal is only long enough to observe the direction of the compass needle. Holding the wire on too long will cause the wire to get hot and discharge the dry cell.
4. Record the direction of the magnetic field as in or out of the paper with reference to Fig. 16.5. Record the answer in Data Table 16.1.

There is a rule, known as the left-hand rule, that is used to determine the direction of a magnetic field around a conductor when the direction of the electron flow is known. The rule states that when the conductor is grasped in the left hand with the thumb pointing in the direction of the electron flow, the fingers will circle the conductor in the direction of the magnetic field. Does this rule agree with your results? If answer is no, contact instructor for assistance.

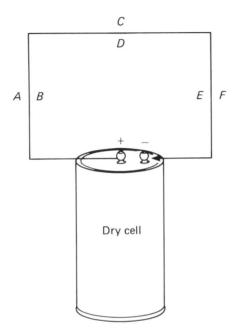

Figure 16.5

QUESTIONS

1. State the general rule for magnetic poles.

2. State Coulomb's law for magnetic poles.

133

3. How is the direction of a magnetic field defined?

4. What is the polarity (north or south) of Earth's magnetic pole located near Earth's north geographic pole? Explain.

5. What is magnetic declination?

6. What is the angle of dip?

7. What is the magnitude of the angle of dip at Earth's magnetic north pole?

8. Explain why the lines of force never cross one another. (*Hint:* Consider the direction of a magnetic line of force.)

Experiment 17

Ohm's Law

INTRODUCTION

The study of electricity is the study of static charges and charges in motion. The flow of electrical charges produces the effect called **electricity**. The electron is the smallest stable particle known, and at rest it possesses the physical property of an electric force field. If the electron is moving, then it possesses, in addition to the electric force field, a magnetic force field. The combined effect of these two force fields is called electricity. The flow of electrons can be brought about by having an excess of electrons at one position and a deficiency at another. If a conducting path is provided between the two positions, the electrons will flow from the position of the excess to that of the deficiency. In an electrical circuit, the electrons flow from the negative terminal (position of excess electrons) through the circuit to the positive terminal (position of deficiency).

Work is done in building up an excess of electrons, therefore they will possess potential energy due to their position. The potential energy per unit charge is known as the **electrical potential energy** and is measured in volts. The rate of flow of charge, called **current**, is measured in amperes. Opposition to the flow of charge is called **resistance** and is measured in ohms. Ohm's law states the relationship between these concepts. The law can be written symbolically as

$$V = IR$$

where V = potential difference, in volts,

I = current, in amperes,

R = resistance, in ohms.

It is important to learn the distinction between rate of flow of electric charge and the amount of flow. The number of electrical charges passing a particular point in a certain period of time is the amount of flow, which is measured in coulombs. The rate of flow is the amount of charge per unit time and is measured in amperes. The relationship is expressed as

$$I = \frac{q}{t}$$

where I =   current (rate of flow), in amperes,

           q =   number of electrical charges (amount of flow), in coulombs,

           t =   time, in seconds.

Ohm's law (expressed in an alternative way) states that the ratio of the voltage (V) to the current (I) in an electrical circuit is a constant and is equal to the resistance (R). This can be written symbolically as

$$\frac{V}{I} = \text{constant} = R$$

The magnitude of an unknown resistance can be determined using this relationship by measuring the voltage across the resistor and the current through it.

LEARNING OBJECTIVES

After completing this experiment, you should be able to do the following:

1. Define the terms voltage, current, and resistance.
2. Determine experimentally the voltage, current, and resistance of an electrical circuit.
3. State Ohm's law.
4. Identify a few basic electrical components and connect them to form an electrical circuit.
5. Use electrical meters to measure voltage and current.

APPARATUS

One 44-ohm rheostat, dc voltmeter (0-10 V), dc milliammeter (0-200 mA), one 40-ohm resistor (5 watt), one unknown resistor, fuse board, connecting wire.

PROCEDURE 1

Connect the 40-ohm resistor and the electrical components in a series circuit, as shown in Fig. 17.1. The voltmeter is connected across (in parallel with) the 40-ohm resistor. **Caution:** Note the polarity when connecting meters. See Fig. 17.1. Have the instructor check the circuit when you have everything connected. He or she will connect the circuit to the low-voltage direct-current power supply. When this has been done, adjust the variable resistor (the rheostat) at different settings and obtain at least six different values for the current and voltage. Proceed with the taking of data by starting with the rheostat at maximum resistance. This will give minimum current in the circuit. Do not allow excess current to flow in the circuit such that any electrical meter goes off scale. Record the measurements in Data Table 17.1.

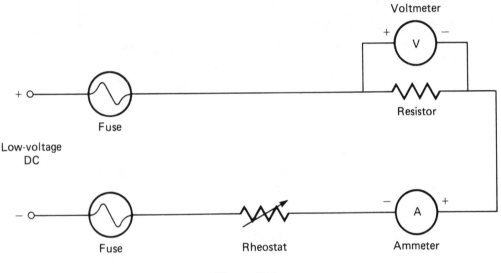

Figure 17.1

PROCEDURE 2

Disconnect the circuit from the supply voltage by removing the plug from the outlet socket. Remove the 40-ohm resistor and replace with the unknown resistor. Have the instructor check the circuit again. He or she will connect the circuit to the low-voltage power supply. Repeat Procedure 1 and record the six voltage and current measurements in Data Table 17.2.

Data Table 17.1   40-ohm Resistor

Rheostat setting	No. 1	No. 2	No. 3	No. 4	No. 5	No. 6
Voltage, in volts						
Current, in amperes						

Data Table 17.2   Unknown Resistor

Rheostat setting	No. 1	No. 2	No. 3	No. 4	No. 5	No. 6
Voltage, in volts						
Current, in amperes						

CALCULATIONS

1. Plot the points and draw the graph for the voltage and current data for the 40-ohm resistor. Plot voltage on the y-axis and current on the x-axis.

   (a) Determine the slope of the curve.   Slope = _____

   (b) How does the value for the slope compare with the magnitude and units of the 40-ohm resistor?

2. Plot the points and draw the graph for the voltage and current data for the unknown resistor. Plot voltage on the y-axis and current on the x-axis.

   (a) Determine the slope of the curve.   Slope = _____

   (b) What is the value of the unknown resistor? R_x = _____

QUESTIONS

1. Connect the circuit as shown in Fig. 17.1 with either resistor, but leave one fuse out of the circuit. This represents a circuit with a burned-out fuse. What is the voltage across the resistor?

   Explain your answer.   _____

2. What is the voltage across the fuse that was left in the circuit?

   Explain your answer.   _____

3. What is the voltage across the terminals of the fuse holder without the fuse?

   Explain your answer.   _____

4. A voltmeter reads 80 volts when connected in parallel with an unknown resistor that has 125 milliamperes flowing through it. What is the resistance of the unknown resistance? Show your work.

                                                    R_x = _____

5. Distinguish between an open and a closed circuit. Give an example of each.

6. A 60-watt light bulb operates on regular house voltage of 120 volts. Normal current through the bulb is 0.5 amp. Determine the resistance of the bulb's filament. Show your work.

                                                    R_x = _____

139

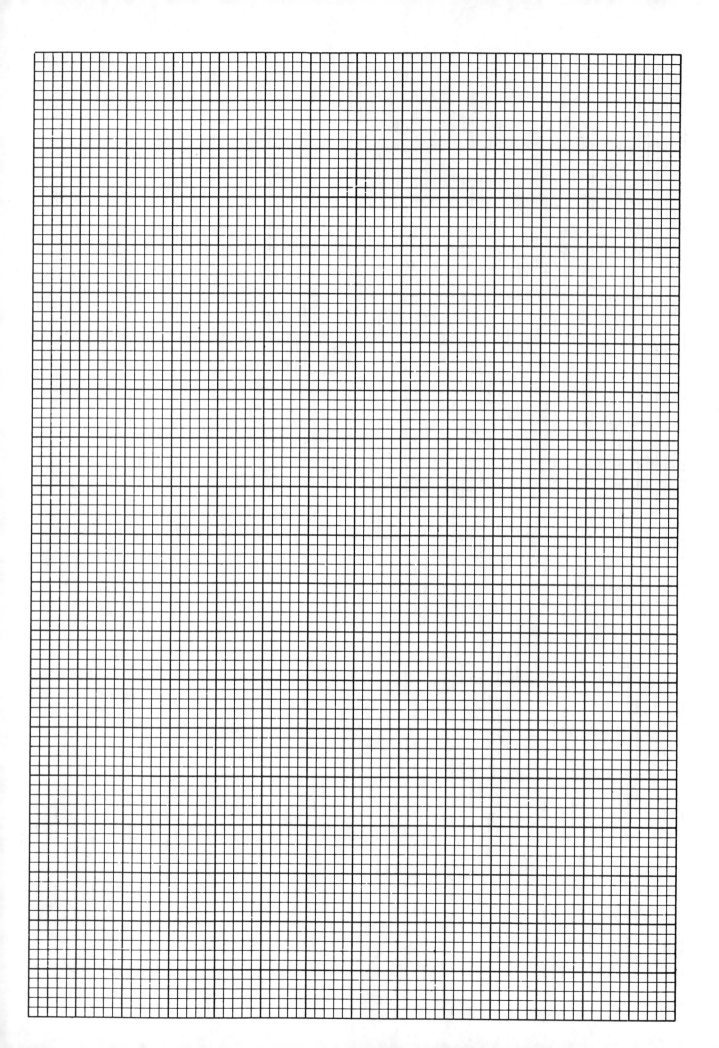

Electromagnetic Waves

INTRODUCTION

When a charged particle such as an electron, proton, or ion is accelerated, energy in the form of electromagnetic radiation emanates in all directions from the accelerated particle. An example of electromagnetic radiation is visible light, which consists of oscillating electric and magnetic fields that travel through vacuum at a speed of 3×10^8 m/s (186,000 mi/s). Your awareness of these printed words is due to the sensitivity of your eyes to electromagnetic radiation, which is reflected from the page of the *Laboratory Guide*.

Electromagnetic radiation consists of oscillating electric and magnetic fields (vector quantities) perpendicular to each other and perpendicular to the direction of the wave motion. The fields are in phase. That is, their minimum and maximum values are at the same points in the oscillating cycle. Fig. 18.1 illustrates the relationship between the force fields and the direction of the wave motion.

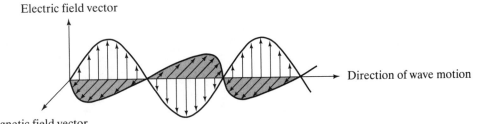

Figure 18.1 *Electromagnetic waves. Electromagnetic waves consist of oscillating electric and magnetic fields. Note that the fields are at right angles to each other and also at right angles to the direction of the wave motion.*

All electromagnetic radiation travels through empty space with a speed c. The speed c, frequency f, and the wavelength λ (lambda) are related by the equation

$$c = f\lambda \tag{18.1}$$

where c = speed of the wave = 3×10^8 m/s,

            f = frequency in hertz (Hz),

            λ = wavelength in meters (m).

Very short wavelengths are expressed in angstrom or nanometer units. One angstrom equals 10^{-10} m; one nanometer equals 10^{-9} m. To convert angstroms to nanometers, move the decimal point one place to the left.

LEARNING OBJECTIVES

After completing this experiment, you should be able to do the following:

1. Explain the physical characteristics of electromagnetic waves.
2. State how electromagnetic waves of different frequencies are generated and detected.
3. Calculate the frequency or wavelength of electromagnetic waves when either the frequency or wavelength is given.
4. Give the relationship between frequency, wavelength, and the speed of electromagnetic waves.

APPARATUS

A hand calculator is useful but not necessary.

PROCEDURE 1

Electromagnetic radiation sensitive to the eye is called visible light and has frequencies ranging from 4.3×10^{14} Hz to 7.5×10^{17} Hz. These are approximate values for the frequencies. Use Eq. 18.1 and calculate the wavelengths (in meters) for the frequencies given in Data Table 18.1. Record your answers in the space provided. Use the conversion information given above and record the wavelengths in angstrom and nanometer units. Use Fig. 18.2 to obtain the color for the calculated wavelengths. Record the color for each wavelength in Data Table 18.1

Data Table 18.1   Calculated wavelengths for visible light frequencies

Frequency f in hertz	Wavelength λ in meters	Wavelength λ in angstroms	Wavelength λ in nanometers	Color
4.3×10^{14}				
5.0×10^{14}				
6.0×10^{14}				
7.5×10^{14}				

The electromagnetic spectrum (an ordered arrangement of frequencies and/or wavelengths) is shown in Fig. 18.2. The spectrum is continuous; it ranges from low frequency radio waves to extremely high frequency gamma rays. There is no sharp dividing line between one type of wave and its neighbor. In fact, there is overlapping of frequencies, as shown in Fig. 18.2. Electromagnetic waves differ from one another only in their frequency or wavelength.

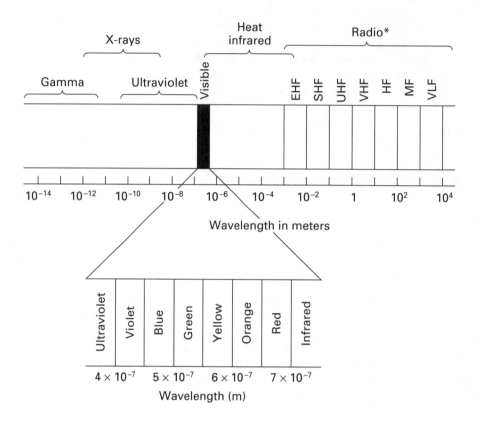

Figure 18.2 *The electromagnetic spectrum.*
* E, extremely; F, frequency; H, high; L, low; M, medium; S, super; U, ultra; V, very.

PROCEDURE 2

Plot a graph that shows the relationship between frequency and wavelength. Use the data in Data Table 18.1. Plot wavelength on the x-axis. (If necessary, refer to Experiment 1 in this *Laboratory Guide* for information on how to plot and label a graph.)

QUESTIONS

1. State in words the relationship between frequency and wavelength that your graph illustrates.

2. Refer to Experiment 1 in the *Laboratory Guide* and give the name that describes the curve in your graph. (*Hint:* What is the equation for the curve?)

PROCEDURE 3

Table 18.2 lists the entire frequency range for each type of electromagnetic radiation. The values for the corresponding wavelengths are not given. Use the equation $c = f\lambda$ and calculate the wavelengths for each of these frequencies. Record the calculated values in the appropriate place in Table 18.2.

Electromagnetic waves are generated and detected in different ways and by different devices, as listed in Table 18.2. Some of these may be unfamiliar to you. It is not feasible to give an explanation for all terms listed in columns one and two of Table 18.2. If you have a question concerning one or more terms, ask your instructor for assistance, or refer to an outside source.

The next paragraph gives a brief explanation of how radio waves are generated by an LC circuit. They are also detected by the same type of circuit.

An LC circuit consists of a capacitor C (two conductors separated by an insulator) and an inductor L (a conductor in the form of a coil). If the capacitor has an initial charge (stored electrons) and the switch S is closed (see Table 18.2, column two), the capacitor will begin to discharge. The total energy in the circuit, which is stored in the electric field between the two conductors, will be transferred by the flow of electrons to the magnetic field of the inductor. The magnetic field cannot sustain itself and collapses. This induces an electron flow in the circuit that places an opposite charge back on the capacitor. The process then repeats itself in the opposite direction. Thus, electrons are accelerated back and forth between the capacitor and the inductor, and electromagnetic waves are generated. Because there is resistance (opposition to the flow of the electrons) in the circuit, energy must be continually supplied to the circuit, or else the oscillations will gradually die out.

PROCEDURE 4

Electromagnetic radiation has a dual nature. Sometimes it acts as a wave, sometimes it acts as a bundle of energy. Max Planck, a German physicist, introduced the idea that a frequency generator could have only discrete or certain amounts of energy. The energy of the generator depends on its frequency according to the relationship

$$E = hf$$

where h (called Planck's constant) $= 6.63 \times 10^{-34}$ (J-s),

 f =   frequency in Hz,

and E =   a quantum of energy (called a photon) in joules.

Calculate the energy of a photon with an associated frequency of (a) 2×10^{14} Hz, (b) 4.3×10^{14} Hz, (c) 7.5×10^{14} Hz, and (d) 3×10^{18} Hz.

147

Table 18.2 Electromagnetic Radiations

Name of Radiation	Radiation generated by	Radiation detected by	Frequency range Hertz (cycles/s) Approximate values	Wavelength range meters To be calculated by student
Radio waves AM broadcast band FM broadcast band TV broadcast band Microwaves	accelerated electrons in an LC circuit 	electronic LC circuit Charged-coupled device (CCD)	AM 0.55×10^6 to 1.60×10^6 FM 88×10^6 to 108×10^6 TV 54×10^6 to 890×10^6 MW 1×10^9 to 1×10^{11}	
Infrared (heat waves)	hot bodies (disturbed molecules)	skin thermometers thermocouples radiometers	1×10^{11} to 4.3×10^{14}	
Visible light	hot bodies (disturbed electrons in atoms and molecules)	eye photographic film photocell	4.3×10^{14} to 7.5×10^{14}	
Ultraviolet	electric arcs special lamps (disturbed electrons in atoms)	photographic film photocell	7.5×10^{14} to 3×10^{17}	
X-rays	special vacuum tubes (disturbed electrons in atoms)	photographic film Geiger counter ionization chamber	3×10^{17} to 3×10^{19}	
Gamma rays	radioactive nuclei (disturbed nucleus of atoms)	Geiger counter ionization chamber scintillation chamber	greater than 3×19^{19}	

QUESTIONS

1. What is the basic explanation for the origin of electromagnetic waves?

2. All electromagnetic waves have the same physical characteristics. They differ only in frequency and wavelength. How are they classified into different types?

3. How do you account for the fact that a sharp dividing line does not exist between one type of electromagnetic wave and its neighbor?

4. How are radio waves generated? How are they detected? What do you think limits the highest radio frequency that can be generated?

5. Why is it impossible to generate radio waves with a frequency of 4×10^{14} Hz?

Use the data given in Table 18.2, Fig. 18.1, and Fig. 18.2 to answer questions 6, 7, and 8.

6. How are X-rays generated? Give three methods for detecting them.

7. What is the minimum frequency generated by radioactive nuclei? The maximum wavelength? What is the name of this radiation?

8. Name the type of radiation that has the greatest range of frequencies. Give the range width.

9. Electromagnetic waves are transverse waves. State the relationship between the electric, magnetic, and velocity vectors.

10. Calculate the difference in energy (joules) between red and violet light.

11. Does a photon associated with blue light have more or less energy than a photon associated with red light? Give the reason for your answer.

Temperature

INTRODUCTION

Temperature can be defined as the degree of hotness or coldness of a body; temperature can also refer to that property of a body that determines the direction of heat flow by conduction. Temperature is also proportional to the average kinetic energy of the random motion of the particles in matter.

Regardless of how temperature is defined, one common way of measuring it is with a mercury-in-glass thermometer. The method of the thermometer's functioning is based on the expansion and contraction effect of heat. Mercury exists in the liquid phase over a large temperature range (melting point, −38.87°C; boiling point, 356.9°C). When a mercury-in-glass thermometer is placed in a gas such as the air, heat is transferred from this gas to the thermometer or from the thermometer to the gas, depending on which is at the higher energy level.

If heat flows from the gas into the glass and the mercury, the glass and the mercury expand to a larger volume and a higher temperature is recorded by the thermometer. If the heat flow is in the reverse direction, the glass and mercury contract and the thermometer records a lower temperature. The coefficient of cubical expansion of mercury is large compared with that of glass; that is, mercury expands much more than glass for each degree change in temperature, so the mercury rises and falls in the glass tube.

A thermometer must be calibrated before a temperature can be recorded accurately. This is done by determining two fixed points (usually the melting point and boiling point of water) on the thermometer, choosing an arbitrary unit of measurement, and marking a scale on the glass bulb.

LEARNING OBJECTIVES

After completing this experiment, you should be able to do the following:

1. Define and explain temperature and state its units of measurement.
2. Calibrate a mercury-in-glass thermometer.
3. Measure temperature with the mercury-in-glass thermometer.

APPARATUS

Nongraduated mercury-in-glass thermometer, 1-L Pyrex glass beaker, tripod base to hold glass beaker, ice cubes, Bunsen burner, marker for making temporary mark on glass, ruler.

PROCEDURE

1. Fill the glass beaker about half full with tap water. Place the thermometer in water and heat the water to boiling point.

2. Allow the system to reach equilibrium; that is, wait until the mercury reaches its highest point in the glass tube and remains there for a few minutes. This is called the steam point.

3. Mark the steam point with the glass marker provided by the instructor. This should be done carefully and quickly.

4. Place a mixture of ice and water in the plastic beaker and position the thermometer in the mixture. Make sure that there is sufficient water in the beaker so that the entire volume of mercury may be completely submerged.

5. Allow the system to come to equilibrium; that is, wait until the mercury reaches its lowest point in the glass tube and remains there for a few minutes. This is called the ice point.

6. Mark the ice point with a glass marker. Make the mark carefully and quickly, and be careful not to break the thermometer in doing so. Be certain that the mark is placed on the glass tube where you observe the ice point.

   Note: Normally, in calibrating a mercury-in-glass thermometer, the ice and steam points are determined with pure water at standard atmospheric pressure. We have not called for either of these conditions for this experiment.

7. Since the expansion of mercury is fairly linear from 0°C to 100°C, a linear scale can be marked on the glass tube over this temperature range. Using the ruler, mark off a scale on the thermometer between the two fixed points. Assign 0° Celsius to the ice point and 100° Celsius to the steam point. Divide into 10° sections and then mark carefully 10 divisions between 20°C and 30°C. This range on the thermometer will be used in determining the existing air temperature.

8. Determine the existing air temperature in the laboratory with your calibrated thermometer.

   _____ °C

9. Ask the instructor for the location in the laboratory of a standard thermometer. Read and record the existing air temperature of the laboratory.

   _____ °C

QUESTIONS

1. What is the least count of your thermometer?   _____

2. What is the percent error of your thermometer in your determination of the existing air temperature in the laboratory? Show your work.

   Percent error...   _____

155

3. Would you expect the steam point on your calibrated thermometer to be higher or lower than a regular standard thermometer? Why?

4. What are the disadvantages of using a water-in-glass thermometer to measure outside air temperature?

5. When a mercury-in-glass thermometer is placed in boiling water, the mercury level falls slightly at first and then begins to rise. Explain this effect. If you have lab time, perform the experiment. Obtain the lab instructor's permission.

Experiment 20

Specific Heat

INTRODUCTION

The law of conservation of energy states that the energy in any isolated system remains constant. This law provides a means for determining the specific heat of a substance.

Experiments have shown that the amount of heat H absorbed or lost by a substance undergoing a temperature change ΔT is directly proportional to the change in the temperature, the mass m of the substance, and the type of substance. We write these relations as follows:

$$H \propto \Delta T \quad \text{and} \quad H \propto m$$

or

$$H \propto m \, \Delta T$$

or

$$H = cm \, \Delta T$$

where H = the heat absorbed or lost in calories,

ΔT = change in temperature in degrees Celsius, $T_2 - T_1$,

m = the mass of the substance in grams,

c = the proportionality constant called the specific heat.

Solving the equation $H = cm \, \Delta T$ for c yields

$$c = \frac{H}{m \, \Delta T}$$

The specific heat of a substance is thus seen to be the amount of heat required to raise a unit mass of a substance one degree in temperature. In terms of cgs units of measurement, the specific heat of a substance is the number of calories required to change the temperature of one gram of the substance one degree Celsius. By definition, the specific heat of water is 1.00 calorie per gram degree Celsius. Because 4.18 joules equals one calorie, the specific heat of water in energy units is 4.18 joules/gram degree Celsius.

In obtaining the experimental value of the specific heat c, it would appear at first to be a simple task to measure the number of calories required to raise one gram of the substance one degree C. True, the mass can be determined by using a balance and the temperature by a thermometer, but we have no method of measuring the amount of heat. Because the heat cannot be directly measured, we resort to a procedure known as the method of mixtures, which is based upon the law of conservation of energy.

The method of mixtures consists of bringing together a known mass of a substance at a known high temperature and a known mass of water at a known low temperature, then determining the resulting temperature of the mixture. In the process, the heat lost by the substance is absorbed by the water and the container holding the water. If we assume that no heat is lost to the surroundings or gained from them, we can express this relation as follows:

Heat lost by substance = heat absorbed by water and container.

The substance in this experiment will be copper or aluminum metal. Therefore, we can formulate the equation

$$H_{\text{metal}} = H_{\text{water}} + H_{\text{container}}$$

Because $H = cm\Delta T$ for any substance, we can substitute for H as follows:

$$(cm\,\Delta T)_{\text{metal}} = (cm\,\Delta T)_{\text{water}} + (cm\,\Delta T)_{\text{container}}$$

or

$$c_{\text{mt}}\,m_{\text{mt}}(T_{\text{mt}} - T_{\text{f}}) = c_{\text{w}}\,m_{\text{w}}(T_{\text{f}} - T_{\text{w}}) + c_{\text{cn}}\,m_{\text{cn}}(T_{\text{f}} - T_{\text{cn}})$$

where
- mt = the symbol for the metal,
- w = the symbol for the water,
- cn = the symbol for the container,
- f = the symbol for final,
- T = temperature in °C,
- m = mass in grams,
- c = specific heat in $\dfrac{\text{cal}}{\text{g}° \cdot \text{C}}$

Solving for the specific heat of the metal c_{mt} we obtain,

$$c_{\text{mt}} = \frac{c_{\text{w}}\,m_{\text{w}}(T_{\text{f}} - T_{\text{w}}) + c_{\text{cn}}\,m_{\text{cn}}(T_{\text{f}} - T_{\text{cn}})}{m_{\text{mt}}(T_{\text{mt}} - T_{\text{f}})} \tag{20.1}$$

All the quantities on the right side of the equals sign are known or can be measured.

LEARNING OBJECTIVES

After completing this experiment, you should be able to do the following:

1. Define specific heat and state its units of measurement.
2. Determine the specific heat of a metal using the method of mixtures.

APPARATUS

Calorimeter, (Cenco or coffee-cup type), steam generator, Bunsen burner, thermometer, balance, aluminum or copper metal pellets.

PROCEDURE

1. Fill the steam generator with tap water to about one-third of its capacity. Light the Bunsen burner and place the flame under the steam generator. Do this first so the water will be heating while you are doing other tasks.
2. Fill the dipper about two-thirds full of the metal. Carefully insert the thermometer in the metal pellets by tipping the dipper to loosen the pellets. Place the dipper in the steam generator and let the water boil until the thermometer reaches its highest temperature and remains constant.
3. While the metal is heating, use the balance to obtain the combined mass of the calorimeter and stirrer. Record the mass in Data Table 20.1.
4. In the introduction to this experiment it was stated that we take a metal at a known *high* temperature and water at a known *low* temperature and bring the two together. The cold water obtained from the cold-water tap will warm to room temperature if allowed to remain standing on the laboratory table. Therefore, do not draw the cold water until the metal has reached and remains constant at its highest temperature. Once the metal has reached this condition, measure and record the temperature of the cold water in the data table. Remove the thermometer from the metal pellets and allow it to cool for use in determining the cold-water temperature. Keep the flame under the steam generator. This will keep the metal at the same high temperature until needed.

    Fill the inner cup of the calorimeter about two-thirds full of cold water. Do not fill the cup full. Remember, the metal is to be added to the cold water. Determine the mass of the cup (with stirrer) plus the water. Record the mass in the data table.
5. The metal and water are now ready to be brought together. With the calorimeter cup positioned in the housing, the thermometer and stirrer placed through the lid of the housing, and the lid slightly removed from the top, pour the metal into the water. Quickly restore the lid and stir the water carefully. Do not lose any water by spilling or splashing. Observe the thermometer closely, and obtain the temperature of the mixture when it has reached its highest equilibrium value. This is the final temperature of the mixture.
6. Remove the thermometer, then determine the combined mass of the calorimeter cup, water, and metal.
7. Complete the data table. Use Eq. 20.1 to calculate the value of the specific heat c_{mt} of the metal.

Table 20.1   Specific Heats

Substance	Specific Heat
Water	$1.00 \ cal/g° \cdot C$
Steam	0.50
Aluminum	0.22
Glass	0.16
Iron	0.105
Copper	0.092

Data Table 20.1

Metal used _____	m_{cn} _____	g
Mass of the calorimeter cup with stirrer (1)	_____	g
Mass of the calorimeter cup + water (2)	_____	g
Mass of the calorimeter cup + water + metal (3)	_____	g
Mass of the water [(2) − (1)]	m_w _____	g
Mass of the metal [(3) − (2)]	m_{mt} _____	g
Temperature of the hot metal	T_{mt} _____	°C
Temperature of the cold water	T_w _____	°C
Temperature of final mixture	T_f _____	°C

CALCULATIONS

1. Determine the specific heat c_{mt} of the metal. Use Eq. 20.1, and show your work. The specific heat of the calorimeter cup (made of aluminum) is 0.22 cal/g°· C.

2. From the accepted value of the specific heat for the metal used in the experiment, determine the percent error. Show your work.

QUESTIONS

1. Distinguish between heat and temperature. Compare definitions and units of measurements.

2. Distinguish between specific heat and heat capacity.

$$\text{Heat capacity} = \frac{H}{\Delta T} = \frac{\text{cal}}{°C}$$

3. State the specific heat of water in terms of energy units (joules).

4. If the final temperature was determined incorrectly (a value greater than the true value was obtained), how would this affect the calculated value of the specific heat? Explain your answer.

5. Where is the greatest possible source of error in this experiment? Why?

Experiment 21

Heat of Fusion

INTRODUCTION

The amount of heat necessary to change one gram of a solid into a liquid (a change of phase) at the same temperature is called the **latent** (hidden) **heat of fusion**. This change of phase is called **melting**. In this experiment the heat of fusion of ice will be determined by using the method of mixtures. Ice, a solid at 0°C will be mixed with water, a liquid, at a temperature of 30 to 35°C.

The law of conservation of energy may be stated in many different ways. For example: The total energy of a system remains constant; energy may be neither created nor destroyed; in changing from one form to another energy is conserved. This experiment, using the method of mixtures to determine the latent heat of fusion of ice, is based upon the law of conservation of energy.

The calorie is the unit of heat (energy) in the metric system. One calorie is defined as the amount of heat necessary to raise the temperature of one gram of pure water one degree Celsius. The specific heat of a substance is the amount of heat, measured in calories, necessary to raise the temperature of one gram of the substance one degree Celsius. Thus, the specific heat of pure water is one calorie per gram degree Celsius.

The definition of specific heat provides an equation for the amount of heat necessary to change the temperature of a given substance. The equation is

$$H = mc(\Delta T) \tag{21.1}$$

where H =   heat gained or lost by the substance,

m =   mass of the substance,

c =   specific heat of the substance,

ΔT =   change in temperature of the substance in °C or K units.

As mentioned above, the phase change from a solid to a liquid is called melting. The heat necessary to change a solid to a liquid is provided by the following equation:

$$H = m H_f \tag{21.2}$$

where H =   heat added to the substance,

m =   mass of the substance,

H_f =   latent heat of fusion of the substance.

Eq. 21.1 and Eq. 21.2 will be used to calculate the heat of fusion of ice. Examine the terms in the equation $H = m H_f$. Notice that in order to calculate H_f, heat of fusion, we must know the value of (m) and the value of (H). The mass (m) we can measure with a balance, but we do not have any method for measuring the heat (H). Since we cannot measure the heat (H), we devise a method to eliminate it from the calculation. This is where the method of mixtures is used.

From the law of conservation of energy we know that when two substances at different temperatures are mixed, the heat lost by one will equal the heat gained by the other.

In this experiment the heat lost by the water and the calorimeter cup holding the water will be equal to the heat gained by the ice plus heat gained by the water produced when the ice melts. This can be stated in words and symbols as follows:

Heat lost by water + heat lost by calorimeter cup = heat gained by ice to melt ice + heat gained by the water from melted ice

$$\text{Heat lost} = \text{Heat gained}$$
$$(mc\,\Delta T)_{\text{water}} + (mc\,\Delta T)_{\text{cup}} = (mH_f)_{\text{ice}} + (mc\Delta T)_{\text{water from melted ice}}$$

Rearranging to solve for H_f

$$H_f = \frac{(mc\,\Delta T)_{\text{water}} + (mc\,\Delta T)_{\text{cup}} - (mc\Delta T)_{\text{water from melted ice}}}{m_{\text{ice}}} \tag{21.3}$$

LEARNING OBJECTIVES

After completing this experiment, you should be able to do the following:

1. Define latent heat of fusion and give an example.
2. State the law of conservation of energy.
3. Give an experimental value for the latent heat of fusion of ice.

APPARATUS

Balance, thermometer (Celsius −10° to 110°), calorimeter, stirring rod, ice cubes, paper towels. *Note*: If a calorimeter is not available, a substitute can be made using a small aluminum can (a soft drink can will do) placed in a Styrofoam cup plus a cardboard lid with holes for the thermometer and the stirring rod.

PROCEDURE

1. Using the balance, determine the mass of the calorimeter cup (inner container) and the stirrer. Record in Data Table 21.1.
2. Fill the calorimeter cup about half full of water at a temperature of approximately 30 to 35°C.
3. Determine the mass of the cup, stirrer plus the water. Record in the data table.
4. Place the calorimeter cup with water and stirrer in the calorimeter housing. Stir the water and determine the temperature to the nearest tenth of a degree. Record in the data table.
5. Dry two ice cubes with paper towels to remove any water, and place them immediately in the calorimeter cup, being careful not to splash any water from the cup.
6. Begin stirring, fairly fast, but be careful not to splash. Stir until all the ice is melted, then determine the temperature to the nearest tenth of a degree. Record in the data table.
7. Remove the calorimeter cup, stirrer, and water from the calorimeter housing and determine their total mass. Record in the data table.
8. Make the calculation necessary to complete the data table.

Data Table 21.1

Mass of calorimeter cup and stirrer	_____	g
Mass of calorimeter cup, stirrer, and water	_____	g
Mass of water	_____	g
Mass of cup, stirrer, water, and melted ice	_____	g
Mass of ice	_____	g
Initial temperature of water, cup, and stirrer	_____	°C
Final temperature of water, cup, and stirrer	_____	°C
ΔT ($T_{initial} - T_{final}$)	_____	°C
Specific heat of water	1 cal/g°· C	
Specific heat of calorimeter cup and stirrer (obtain from instructor)	_____	cal/g°· C
H_f, heat of fusion of ice (experimental value)	_____	cal/g

9. Calculate the value for the heat of fusion of ice using Eq. 21.3. Show your work. Record the value of H_f in the data table.

CALCULATIONS AND QUESTIONS

1. The accepted value for the heat of fusion of ice is 80 cal/g. This is the amount of energy that must be transferred into one gram of ice at its melting temperature (0° C) in order to change it into liquid water. Also, 80 cal/g is the amount of energy that must be removed from one gram of water at 0° C in order to change liquid water to ice. Calculate the percent error using this value as the standard. Show your work.

2. How will using wet ice affect the experimental value for the latent heat of fusion of ice? Explain.

3. If the calorimeter is not insulated, how will heat from the air in the laboratory affect the experimental value for the latent heat of fusion of ice? Explain.

4. How do citrus growers in Florida use heat of fusion to protect against frost damage?

5. Is it possible to add heat energy to a substance and not increase its temperature? Give an example.

Heat of Vaporization of Water

INTRODUCTION

The amount of heat necessary to change one gram of vapor into a liquid (a change in phase) at the same temperature is called the latent (hidden) *heat of vaporization*. This change of phase is called *condensation*. In this experiment, the heat of vaporization of water will be determined by using the method of mixtures. Steam, a vapor, at 100°C will be added to water, a liquid, at a temperature of 15°C to 25°C.

The law of conservation of energy may be stated in many different ways. For example: The total energy of a system remains constant; energy may be neither created nor destroyed; in changing from one form to another, energy is conserved. This experiment, using the method of mixtures to determine the latent heat of vaporization of water, is based upon the law of conservation of energy.

The calorie is the unit of heat (energy) in the metric system. One calorie is defined as the amount of heat necessary to raise the temperature of one gram of pure water one degree Celsius. The specific heat of a substance is the amount of heat, measured in calories, necessary to raise the temperature of one gram of the substance one degree Celsius. Thus, the specific heat of pure water is one calorie per gram degree Celsius.

The definition of specific heat provides an equation for the amount of heat necessary to change the temperature of a given substance. The equation is

$$H = mc(\Delta T) \tag{22.1}$$

where H = heat gained or lost by the substance,

         m = mass of the substance,

         c = specific heat of the substance,

       ΔT = change in temperature of the substance in °C or K units.

As mentioned above, in the method of mixtures, the law of conservation of energy requires that the heat given up by the steam be equal to the heat gained by the water. The heat given up by the steam in condensing into water is provided by the following equation:

$$H = m_s H_v \tag{22.2}$$

where H = the heat given up by the steam,

           m_s = mass of the steam,

           H_v = the latent heat of vaporization.

After the steam condenses into water, the mass (m_s) will continue to decrease in temperature due to mixing with the cool (15°C to 25°C) water in the calorimeter cup. The complete process can be stated as follows:

Heat given up = Heat absorbed

$$\begin{bmatrix} \text{condensation} \\ \text{of steam} \end{bmatrix} + \begin{bmatrix} \text{cooling of} \\ \text{condensed steam} \end{bmatrix} = \begin{bmatrix} \text{warming up of cool water in} \\ \text{calorimeter cup plus the cup} \end{bmatrix}$$

$$(mH_v)_{\text{steam}} \quad + \quad (mc\Delta T)_{\text{water}} \quad = \quad (mc\Delta T)_{\text{water}} + (mc\Delta T)_{\text{cup}}$$

$$m_s H_v \quad + \quad m_s c_w(T_s - T_f) \quad = \quad m_w c_w(T_f - T_w) + m_{\text{cup}} c_{\text{cup}}(T_f - T_w)$$

Rearranging to solve for H_v yields

$$H_v = \frac{m_w c_w(T_f - T_w) + m_{\text{cup}} c_{\text{cup}}(T_f - T_w) - m_s c_w(T_s - T_f)}{m_s} \tag{22.3}$$

Where the symbols are defined in Table 22.1.

(*Note:* Here T_s is the average temperature in degrees Celsius of steam at the elevation where the experiment is performed. The steam point temperature varies with altitude above sea level. The value for T_s is usually about 99°C in most cases.)

LEARNING OBJECTIVES

After completing this experiment, you should be able to do the following:

1. Define latent heat of vaporization and give an example.
2. State the law of conservation of energy.
3. Give an experimental value for the latent heat of vaporization of water.
4. Explain the method of mixtures and explain how the method can be used to determine the latent heat of vaporization.

APPARATUS

Balance, boiler with hose and steam trap, boiler stand, ring stand, clamp, Bunsen burner or other heat source (an electric boiler unit may be used), small beaker, calorimeter, stirring rod, a thermometer (−20°C to 110°C).

PROCEDURE

1. Assemble the apparatus as shown in Fig. 22.1. Keep the steam outlet hose in the beaker instead of in the calorimeter.
2. Fill the boiler about one-third full of water and place it over the heat source so that the water will heat and boil. While you are waiting for the water to boil, complete steps 3–6.
3. Use the balance and determine the mass of the calorimeter cup (inner container) and the stirrer. Record in Data Table 22.1 on the next page.
4. Fill the calorimeter cup about half full of water at a temperature of approximately 15°C to 25°C.
5. Determine the mass of the calorimeter cup, stirrer, and the water. Record in the data table.
6. Place the calorimeter cup with water and stirrer in the calorimeter housing. Stir the water and determine the temperature to the nearest tenth of a degree. Record in the data table.
7. When a heavy flow of steam is coming from the outlet base, carefully place the end of the delivery hose in the inner container of the calorimeter through the hole in the top. Make sure the end of the delivery hose is submerged so that the steam will bubble through the water already in the cup. The thermometer should also be in the cup while steam is being added. (Note: boil the water gently so that excessive pressure does not build up in the boiler or hoses.)

Figure 22.1 *Apparatus for determining the heat of vaporization.*

Data Table 22.1

No.	Description		
1	Mass of calorimeter cup and stirrer	m_{cup}	_____ g
2	Mass of calorimeter cup, stirrer, and water		_____ g
3	Mass of water (2–1)	m_w	_____ g
4	Mass of cup, stirrer, water, and condensed steam		_____ g
5	Mass of steam (4–2)	m_s	_____ g
6	Initial temperature of water, cup, and stirrer	T_w	_____ °C
7	Final temperature of water, cup, and stirrer	T_f	_____ °C
8	$\Delta T\ (T_f - T_w)$		_____ °C
9	Specific heat of water	c_w	1 cal/g°C
10	Specific heat of calorimeter cup and stirrer (obtain from instructor)	c_{cup}	_____ cal/g°C
11	Temperature of steam	T_s	_____ °C
12	Heat of vaporization of water (experimental value)	H_v	_____ cal/g

8. Watch the temperature; it will go up quite rapidly. As soon as the thermometer reads 40°C–45°C, remove the steam hose. Do not shut off the heat until after the hose has been removed from the water.

9. Stir the water carefully, then determine the temperature to the nearest tenth of a degree. Record in the data table.

10. Remove the calorimeter cup, stirrer, and water from the calorimeter housing and determine their total mass. Record in the data table.

11. Make the calculations necessary to complete the data table.

12. Calculate the value for the heat of vaporization of water using Eq. 22.3. The final temperature of the condensed steam will be the same as the final temperature of the water in the calorimeter cup.

CALCULATIONS AND QUESTIONS

1. The accepted value for the latent heat of vaporization is 540 cal/g. This is the amount of energy that must be removed from one gram of steam at its boiling temperature (100°C) in order to change it into liquid water. Also, 540 cal/g is the amount of energy that must be added to one gram of water at 100°C in order to change liquid water to steam. Calculate the percentage error in your experimental value of H_v, using this value as the standard. Show your work.

2. How will the experimental value be affected if some water (liquid) comes through the hose into the calorimeter cup along with the steam?

3. If the calorimeter is not insulated, how will heat from the air in the laboratory affect the experimental value for the latent heat of vaporization?

4. How does a steam turbine use the heat of vaporization to increase the output of electrical energy it is generating?

5. Is it possible to remove heat energy from a substance and not change its temperature? Explain, and give an example.

6. Calculate the percentage increase in the mass of the water in the calorimeter cup with the addition of the condensed steam.

Inverse-Square Law of Radiation

INTRODUCTION

A radioactive substance emits, at random, radiation in the form of matter, called **alpha** or **beta particles**, and electromagnetic waves, called **gamma rays**. Because the radiation is independent of temperature, pressure, chemical state, electric fields, magnetic fields, and other external conditions, the radiation is considered to be a property of the nucleus of the atom.

If the substance is considered a point source and the radiation is emitted equally in all directions, the number of particles per unit area will vary inversely as the square of the distance from the source.

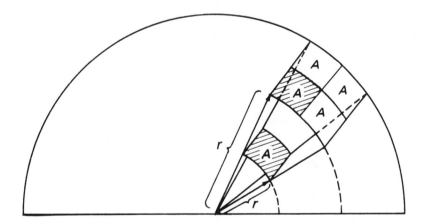

Figure 23.1

Fig. 23.1 is a diagram illustrating the geometry of a sphere. The area of a sphere may be determined by the equation

$$A = 4\pi r^2$$

where A = area and r = radius of the sphere.

In a sphere with a radius of one meter,

$$A = 4\pi \cdot 1^2$$

and the radiation coming from the source must pass through this area.

In a sphere with a radius of two meters,

$$A = 4\pi \cdot 2^2$$

and the area (expressed in square meters) is four times the first area. Because the number of particles that must pass through this larger area remains the same per unit area, the number of particles is one-fourth as much. In other words, when we double the distance, the particle intensity, which is the number of particles passing through or falling upon unit area in unit time, is one-fourth the original value:

$$I = \frac{N}{4\pi s^2}$$

where I = particle intensity,

            N = total number of particles emitted per unit time,

            s = distance from source to point of intensity measurement.

LEARNING OBJECTIVES

After completing this experiment, you should be able to do the following:

1. State the inverse-square law of radiation.
2. Take radiation measurements with electronic equipment.
3. Determine experimentally the inverse-square law of radiation.

APPARATUS

Source of gamma radiation, detector with power supply, timer, meter stick.

PROCEDURE

The electronic detector, associated power supply, and measuring device will be set up and explained by the laboratory instructor.

The gamma source will be positioned on the meter stick for the initial reading. The students will take turns reading the intensity at a large number of positions as the source is moved away from the detector. Record the intensity and distance in Data Table 23.1.

CALCULATIONS

1. Calculate $1/s^2$ for each distance.

2. Plot intensity versus $1/s^2$ and draw a straight line for the best fit.

QUESTIONS

1. What shape of curve is obtained if intensity is plotted against distance?

2. What two force laws previously studied obey the inverse-square law?

3. What are gamma rays? Describe their features.

4. At a distance of 1 m from a radioactive source, the intensity is I. What is the intensity at a distance of 3 m?

Data Table 23.1

Distance s, in cm	Intensity I, in counts per minute	$1/s^2$

The Rutherford Scattering Box

INTRODUCTION

This experiment is designed to give the student an insight into how scientists work. In investigations of matter, for example, its microscopic structure is experimentally determined by using some mechanism that interacts with the matter. This mechanism acts as a probing "tool," and from a knowledge of its properties and the monitored results of its interaction with matter, the scientist forms a model of the structure of the matter that would explain these results.

An example of such a tool is the particle accelerator used to probe nuclear structure, such as the 11-million-electron-volt Tandem Van de Graaff Accelerator. An accelerator is a machine that accelerates electrically charged particles (electrons, protons, or ions) to high velocities so that they have large kinetic energies. With the Tandem Van de Graaff generator, energies up to several million electron volts (1 eV = 1.69×10^{-19} joule) can be obtained.

In the accelerator, negative ions such as H^- (an atom containing one proton and two electrons) are accelerated across a potential difference toward a generator. At the generator, they collide with a gas and the electrons of the ions are stripped off. The resulting positively charged protons are then accelerated away from the generator across the reverse potential drop. The effect is that of having two accelerators in a line, hence the name "tandem." The beam of accelerated particles is then directed toward a target material composed of the nuclei that are being investigated. See Fig. 24.1 for a schematic diagram of a Tandem Van de Graaff Accelerator.

One aspect of nuclear structure may be investigated by observing the scattering of the probe particles from the target nuclei. Lord Rutherford first used a particle probe (alpha particles) from a natural radioactive source to show that the protons of an atom were concentrated at the center of the atom as a nucleus. This model was deduced as being consistent with the observed scattering of the alpha particles.

Figure 24.1 *Schematic diagram of a Tandem Van de Graaff Accelerator.*

LEARNING OBJECTIVES

After completing this experiment, you should be able to do the following:

1. Determine experimentally the shape of an unknown target (object) by probing with a projectile.
2. Apply the deductive process of reaching a conclusion.

APPARATUS

Several Rutherford scattering boxes with different wood targets of unknown shapes are used. Each box is equipped with a quantity of probe particles (marbles) and a mechanical accelerator.

See Fig. 24.2 for a photograph of the Rutherford Scattering Box. The box can be made from plywood or other suitable material. The mechanical accelerator is a small section of bent tubular pipe large enough for the marbles to pass through. The targets are cut from solid wood with a thickness slightly greater than the diameter of the marbles. The target is hidden from the experimenter's view by the target zone shield that fits over the target and is held in place by a screw. This shield is not to be removed during the experiment. The experimenter must use another Rutherford Scattering Box to work with a different target.

PROCEDURE

By probing the target with the marble particles and observing the scattering, the student is to determine the shape of the target. The target may be probed from any angle. Probing is done by allowing the marble to roll down the bent tube that is pointed at the target. Reference lines have been drawn on the bottom of the box to assist in orientation.

Use a different diagram for each scattering box, and draw lines to indicate the paths the marbles trace when probing the targets. When you have determined the shape of the target, draw the target shape in the space provided. The shapes of the targets are made up of straight and curved surfaces. Think of how an incident particle would be scattered from a given surface and how you could confirm that shape of this surface by slightly varying the incident angle. Remember the probing must be made completely around the target to determine its complete shape.

Figure 24.2   *Rutherford Scattering Box.*

QUESTIONS

1. What is the polarity (plus or minus) of the high-voltage terminal of the Van de Graaff generator if it accelerates the H$^-$ ions as described in the introduction?

2. How can a beam of accelerated charged particles be maneuvered and directed toward targets in different locations? The accelerator shown in Fig. 24.1 has two target areas at right angles to the initial particle beam. The beam may be directed into either area.

3. What was the force of interaction between the probe particles and the nucleus that caused the particles to be scattered and led Lord Rutherford to deduce the atom had a "core," or nucleus, of protons? (Prior to this experiment it was thought that the atom had protons distributed throughout its structure.)

185

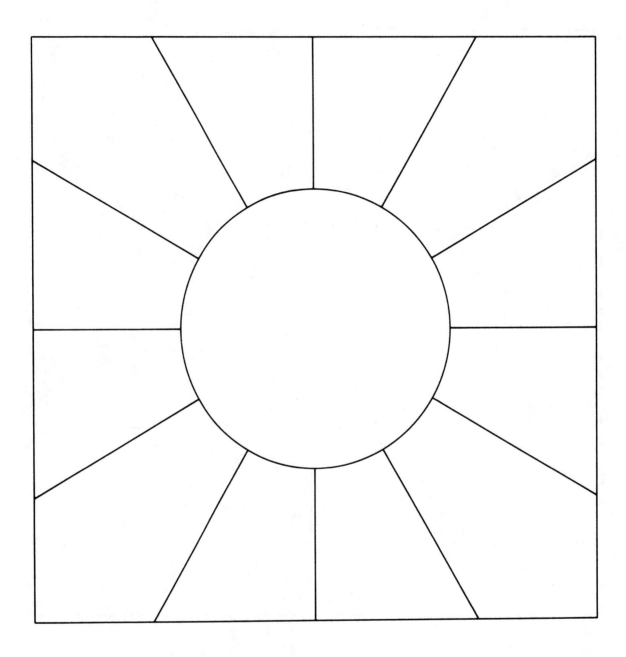

Radiation

INTRODUCTION

There are three main kinds of radiation that might be radiated by a radioactive substance. The emitted radiations are alpha particles (helium-4 nuclei), beta particles (electrons), and gamma rays (high-energy, high-frequency electromagnetic radiation). These three kinds of radiation have different properties. For instance, the gamma rays are more penetrating, and the alpha particles and electrons are deflected by magnetic fields.

In our day-to-day life we are constantly exposed to natural radiation. This is radiation due to cosmic rays bombarding our atmosphere and creating radioactive substances, or due to the natural decay of radioactive substances present in Earth. There are also unnatural radioactive substances due to nuclear explosions conducted in the atmosphere.

The intensity of the radiation received from a radioactive source varies as the inverse square of the distance from the source. Thus, if the distance from the source is doubled, the radiation received by a given area is reduced by one-fourth.

LEARNING OBJECTIVES

After completing this experiment, you should be able to do the following:

1. State the properties of alpha, beta, and gamma radiation.
2. Determine experimentally some of the properties of radiation.

APPARATUS

Radiation sources, detector with power supply, timer, meter stick, magnets, lead shielding.

PROCEDURE

1. First investigate the background radiation. With all radioactive sources far removed from the detector, determine the number of counts for five minutes. Do this three times.

   Compute the counts/min.   Background radiation: #1 _____

                                                           #2 _____

                                                           #3 _____

                                            Average background _____

2. Next examine radioactive source number one. Determine the intensity of the radiation in counts per minute produced by the source for four different distances from the detector. Record your data. Plot the intensity versus the distance on your *graph paper*.

Intensity (counts/minute)	Distance from source (s)	$\frac{1}{d^2}$
1		
2		
3		
4		

3. Examine the penetrating effects of rays from the different sources. Measure the radiation produced, first with no shielding and then with a sheet of paper, a piece of wood, and a lead shield.

Source	Distance	Intensity (counts/minute)			
		No shielding	Paper	Wood	Lead shield
1					
2					

4. Examine the effects of the different sources when the radiation travels through a magnetic field.

Source	Distance	Intensity (counts/minute)	
		With no magnetic field	With a magnetic field
1			
2			

5. Alpha particles have a charge of +2 and are about 7500 times more massive than beta particles. Beta particles have a charge of −1. Gamma rays have no mass or charge. To detect alpha particles, a special detector is needed, because they are easily absorbed by matter. Your detector would absorb the alpha particles before they could arrive at the sensitive part of the detector.

From the properties of beta particles and gamma rays, determine what kind of radiation is given off by sources 1 and 2.

Source 1 _____

Source 2 _____

QUESTIONS

1. Why aren't the three readings in Procedure 1 identical?

2. Is there a relation between $1/d^2$ and the intensity? If so, what? Should there be a relationship? Discuss sources of error in this part of the experiment.

3. Why are alpha particles so easily absorbed?

Experiment 26

Spectroscopy

INTRODUCTION

A sodium vapor lamp contains a gas at low pressure. Under this condition of low pressure, the gas is excited by fast-moving electrons from the negative electrode. When the energy transferred to an atom is sufficient, an outer (or valence) electron is excited, and the atom changes from its normal state of energy E_1 to the first excited state of energy E_2, which is greater than E_1. The electron in the excited state then radiates energy and subsequently returns to its normal state. The frequency of the radiation is related to the energy difference by the following relationship,

$$hf = E_2 - E_1 \qquad (26.1)$$

where E_1 = energy of the electron in its normal state, in joules,

         E_2 = energy of the electron in its excited state, in joules,

         f = frequency of the radiation in cycles per second,

         h = Planck's constant, which has a value of 6.6×10^{-34} joule · second.

The relationship between the velocity c of electromagnetic radiation, the wavelength λ, and the frequency f is

$$f = \frac{c}{\lambda}$$

Substituting in Eq. 26.1, we obtain

$$\frac{hc}{\lambda} = E_2 - E_1$$

The difference between the two energies can be obtained if λ is measured, since h and c are known constants.

Many different frequencies or wavelengths are generated from an electrical discharge because there are many energy levels in an atom and the electrons may be transferred to any of these levels. Also, the electron may return to its normal state through a series of transitions rather than through one transition.

When the radiation is allowed to fall on a diffraction grating, the frequencies or wavelengths are resolved into their individual components. A diffraction grating is made by cutting on the surface of a piece of glass very fine, narrow, parallel grooves spaced very close together. Where the grooves are cut, the glass will be opaque, but the spaces between grooves will allow the radiation to pass through. Some gratings are made with as many as 10,000 spaces or lines per centimeter. Each of the narrow slit openings acts as a source of radiation and the radiation from each opening interferes with that from all the other openings. Constructive interference occurs along a line perpendicular to the plane of the grating and also along lines that are at angles with the perpendicular, according to the following equation:

$$n\lambda = d \sin \theta \qquad (26.2)$$

where λ = wavelength,

d = distance between any two grooves of the grating, known as the grating space,

θ = angle of diffraction measured from the normal,

n = an integer (any whole number, not zero).

If $n = 1$, an image will appear at some angle. For $n = 2, 3, 4, \ldots$, a similar image will appear at greater angles. These images are called orders of the spectrum. See Fig. 26.1.

Figure 26.1

LEARNING OBJECTIVES

After completing this experiment, you should be able to do the following:

1. State the relationship between velocity, frequency, and wavelength of electromagnetic radiation.
2. Determine theoretically and experimentally the wavelength of light produced by a sodium vapor lamp.
3. Calculate the energy necessary to produce the radiation emitted by the sodium atom.

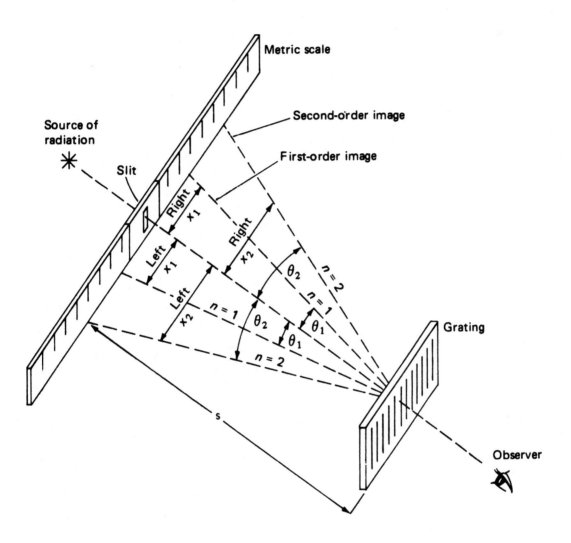

Metric scale

Second-order image

Source of
radiation

First-order image

Slit

Right x_1

Right x_2

Left x_1

Left x_2

θ_2

$n = 2$

$n = 1$

θ_1

$n = 1$

θ_2

θ_1

Grating

$n = 2$

s

Observer

Figure 26.2

APPARATUS

Diffraction grating, grating holder, sodium vapor lamp or mercury arc lamp, two meter sticks, two supports for meter sticks, slit with attached scale (Cenco 86260-2).

199

PROCEDURE

1. Plug in the sodium vapor lamp and adjust its height so that the lamp is centered in front of the slit.
2. Position the sodium lamp, the slit, and the grating in a straight line, with the plane of the grating perpendicular to the rays of light from the slit. Observe the pattern with a distance s between the slit and the screen of 40 cm, 60 cm, 80 cm, and 100 cm. Note the difference between the pattern positions x_1 and x_2 for the first two orders in each case. See Fig. 26.2.
3. Look through the grating and observe the diffracted images on both sides of the center line. If the grating space d is 2.54/7500 cm, you will see two yellow lines plus other colored lines of the spectrum. If the grating space d is 2.54/600 cm, only one yellow line will appear. The 600-lines-per-inch grating cannot resolve the two yellow lines.

   The images of the slit should appear at equal distance from the center line. If they do not, rotate the grating slightly until they do.
4. Set the distance d for the various values given in Data Table 26.2. Measure the angle θ for the yellow line in the first order on each side of the center line and record values in Data Table 26.2. An approximate value of sin θ can be obtained by measuring the apparent displacement (in cm) of the yellow lines of the sodium spectrum and dividing this displacement by the distance d. Always measure to the first-order ($n = 1$) lines for gratings having 7500 lines per inch and to the fifth-order ($n = 5$) lines for gratings having 600 lines per inch. See Fig. 26.2.

Data Table 26.1

Type of lamp used	_____
Lines per inch in grating	_____
Grating space d	_____ cm

If the grating has 7500 lines or more per inch, measure the displacement x_1 and x_2 for $n = 1$. If the grating has 600 lines per inch, measure the displacement x_1 and x_2 for $n = 5$.

Data Table 26.2

Distance d between grating and slit, in cm	x_1 Left	x_1 Right	Average of x_1 Left and x_1 Right	x_2 Left	x_2 Right	Average of x_2 Left and x_2 Right	*Sin θ_1	*Sin θ_2
40								
60								
80								
100								
Average values of sin θ								

*For any value of sin θ: $\text{Sin } \theta = \dfrac{\text{side opposite angle } \theta \text{ (use average value)}}{\text{hypotenuse (use distance } d)}$

Determine the average values of x_1 right and x_1 left, and the average values for x_2 right and x_2 left. Then calculate θ_1 and θ_2 and determine their average values.

CALCULATIONS

1. Using Eq. 26.2, solve for the average wavelength of the yellow lines of sodium.* Accepted value is 5893 angstroms. (*Note*: 1 angstrom = 1×10^{-8} cm.) For $n = 1$ use average value of $\sin \theta_1$. For $n = 2$ use $\sin \theta_2$.

2. Use $hc/\lambda = \Delta E$ to calculate ΔE, the energy of the first excited state.

---

*Prominent mercury spectral lines in angstroms: violet—4358; green—5461; red—6907.

EXPERIMENT 26   NAME (print) _____ DATE _____
                          LAST   FIRST
LABORATORY SECTION _____ PARTNER(S) _____

QUESTIONS

1. What is a diffraction grating?

2. What is the grating space?

3. In the first-order spectrum, is the red line or the violet line closest to the central image?

Density of Liquids and Solids

INTRODUCTION

Density is a physical property of matter. It is a derived quantity that is defined as the mass per unit volume. In symbolic notation:

$$\text{density} = \frac{\text{mass}}{\text{volume}} = \frac{\text{mass}}{(\text{length})^3} = \frac{m}{l^3} \tag{27.1}$$

In SI units, density is measured in kg/m^3. Other units of density are: g/cm^3 or g/mL, and lb/ft^3. In this experiment units of g/cm^3 or g/mL will be used. This is because most laboratory equipment is still calibrated in cgs units (that is, centimeters, grams, and seconds), even though most scientific texts are now using the SI system of units.

The maximum density of pure, air-free water at atmospheric pressure, which occurs at 3.98°C, is defined to be exactly 1.0000 g/cm^3. As the temperature of water increases above or decreases below 3.98°C, its density decreases. Table 27.1 gives the density of pure water as a function of temperature in the region of 3.98°C and near room temperature (15°C–25°C).

Data Table 27.1   Density of Pure Air-Free Water at Atmospheric Pressure

Temperature °C	Density gm/mL	Temperature °C	Density gm/mL
0	0.99987	24	0.99733
3.98	1.00000	25	0.99708
10	0.99973	26	0.99681
15	0.99913	27	0.99654
20	0.99823	28	0.99626
21	0.99802	29	0.99598
22	0.99780	30	0.99568
23	0.99757		

Other substances can have densities less than or greater than that of pure water. For example, the most dense solid known is osmium, which has a density of 22.5 g/cm³, while gasoline (a liquid) has a density of 0.68 g/cm³, and balsa wood (a solid) has a density of only 0.12 g/cm³. As you can see, the phase of a sample of material is not always a good indication of its density.

The density of any substance can be determined if we can accurately determine its mass and volume. The mass of a solid or liquid can be measured on a pan balance, which allows us to compare the mass of the unknown material with a standard set of masses that are experiencing the same gravitational force. You might also use a triple-beam balance, on which the positioning of known masses along a preset scale will indicate the mass of the sample under study.

The volume of a liquid can be found with a precision of about ±1 ml by using a graduated cylinder. A more precise determination can be made using a pycnometer (pik'-nom'-ə-t ə r), which is a bottle-like container having a fixed definite volume. Both procedures will be explained in this experiment, and either or both can be performed, depending on the time and equipment available.

The volume of a solid that has a regular shape can be determined by directly measuring its dimensions. The volume can then be calculated using one of the equations given in Table 27.2. In many instances it is not possible to measure the dimensions of an irregularly shaped solid, and an indirect method must be used. One such method is to measure the volume of water (or other liquid) that is displaced when a solid sample is completely immersed in liquid. If the displaced liquid is captured in a graduated cylinder, its volume can then be determined. The volume of the displaced liquid will be equal to the volume of the submerged solid object. It is also possible to completely submerge the sample in a partially filled graduated cylinder. The difference in volumes before and after the material is submerged will give the volume of the sample.

LEARNING OBJECTIVES

After completing this experiment, you should be able to do the following:

1. Define and give an example of density.
2. Explain how the density of a solid or a liquid can be calculated.
3. Experimentally determine the density of any small sample of liquid or solid using simple laboratory equipment.

APPARATUS

Laboratory balance with a sensitivity of at least 0.01 grams (0.001 grams is preferred), pycnometer (volumetric flask with stopper), graduated cylinder (100 mL or larger), thermometer, ruler or meter stick, calipers or micrometers, water (distilled, if possible), an "unknown" liquid, an "unknown" regularly shaped solid material, such as a block, cylinder, or sphere, an irregular shaped sample, such as a rock or a piece of metal, and a bulk sample (aluminum shot, copper shot, lead shot, or small glass marbles. For best accuracy, these samples need to be small enough to be placed in the volumetric flask or the graduated cylinder.

Data Table 27.2   Equations for Calculating Volume

Volume of a BLOCK	length × width × height
Volume of a CYLINDER	$\pi \times radius^2 \times height$
Volume of a SPHERE	$4/3 \times \pi \times radius^3$
$\pi = 3.1416$	

PROCEDURE 1

The purpose of this procedure is to determine the exact volume of the volumetric flask.

1. Weigh the clean dry flask and stopper on the balance. Record this mass, to the nearest 0.001 g if possible, in the data table.
2. Determine the temperature of the water. Ideally, the water temperature should be the same as the air temperature in the laboratory. Record this temperature in the data table.
3. Use Table 27.1 to obtain the exact density of the water at the temperature that you previously measured. Record this density in the data table.
4. Fill the flask with water and *carefully* insert the stopper to remove excess water and any air bubbles. Make sure the stopper is seated properly in the flask and that the flask and stopper neck are completely filled with water.
5. Using a dry paper towel, carefully remove any water that may be on the stopper or on the outside of the flask.
6. Weigh the flask + stopper + water to the nearest 0.001 g, if possible. Record this mass in the data table.
7. Determine the mass of the water in the flask by subtracting:

$$(\text{mass of flask} + \text{stopper} + \text{water}) - (\text{mass of flask} + \text{stopper})$$

8. Calculate the exact volume of the flask using Eq. 27.1. Be careful to round off your answers to the appropriate number of significant digits. Show your calculations on a separate sheet of paper. Record this volume in the data table.

PROCEDURE 2

The purpose of this procedure is to determine the density of an unknown liquid.

1. Empty the water from the flask. Dry the flask and stopper thoroughly. Ask the instructor where to empty the water.
2. Fill the flask with an unknown liquid.
3. Carefully place the stopper in the flask. Dry the outside of the stopper and flask thoroughly.
4. Weigh the flask + stopper + unknown liquid and record the mass, to the nearest 0.001 g, if possible, in the data table.
5. Calculate the mass of the unknown liquid. Record this mass in the data table. Show your calculations on a separate sheet of paper. This mass may be found by subtracting:

$$(\text{mass of flask} + \text{stopper} + \text{liquid}) - (\text{mass of flask} + \text{stopper})$$

6. Calculate the density of the unknown liquid using Eq. 27.1. Record the density in the data table. Show your calculations on a separate sheet of paper. Be sure all calculations contain the correct number of significant digits.

PROCEDURE 3

The purpose of this procedure is to determine the density of an irregularly shaped unknown solid.

1. Empty the unknown liquid from the flask and wash and dry the flask and stopper. Ask the instructor where to empty the unknown liquid.
2. Place 50–60 grams of the bulk solid sample in the volumetric flask. Make sure the solid is clean and dry.
3. Weigh the flask + stopper + unknown solid. Record the mass, to the nearest 0.001 g if possible, in the data table.

4. Add water to the flask immersing the solid. Agitate the solid slightly to remove any air bubbles, then fill the flask completely and insert the stopper. Observe closely for any air bubbles in the flask. If any bubbles are still present, remove the stopper, add water, and replace stopper.
5. Weigh the flask + stopper + water + solid. Record the mass, to the nearest 0.001 g, if possible, in the data table.
6. Calculate the mass of the unknown solid. Record this mass in the data table. Show your calculations on a separate sheet of paper. This mass can be determined by subtraction:

(mass of flask + stopper + unknown solid) – (mass of flask + stopper)

7. Determine the volume of the unknown solid. The volume is equal to the volume of water displaced. Record the volume of the unknown solid in the data table. This volume can also be determined by subtraction:

(volume of flask) – (volume of water added on top of unknown solid)

The volume of the flask was found in Procedure 1. The volume of the water is a little trickier to find. The mass of water added on top of the unknown solid can be found by subtraction:

(mass of flask + stopper + water + solid) – (mass of flask + stopper + unknown solid)

The volume of water is then the mass of water times the density of water at your working temperature.
8. Calculate the density of the unknown solid. Record the density in the data table. Show your calculations on a separate sheet of paper. Be sure all calculations contain the correct number of significant digits.

PROCEDURE 4

The purpose of this procedure is to determine the density of an unknown solid using the displaced-water method.

This is a less-accurate method for obtaining the density of a solid, but it may be helpful in showing the student a more direct method of determining volume. This procedure may also be used when volumetric flasks are not available, in which case the experiment should be repeated several times.

1. Determine the mass of an unknown solid by using a laboratory balance. Record this mass in the data table.
2. Fill the graduated cylinder approximately half full of water. Carefully read the level of the water in the graduated cylinder and record this value in the data table. Attach a piece of fine thread (25 to 30 cm long) to the solid sample. Lower the sample into the graduated cylinder until the sample is completely submerged. Read the new level of the water in the graduated cylinder, and record this value in the data table. The difference between the two readings will be the volume of displaced water and is, therefore, exactly equal to the volume of the solid. Record this volume in the data table.

If the solid sample happens to float in the water, you will have to use a long thin rod, such as a pencil, to force the sample completely under water, or your determination of density will not be correct. Any portion of the rod that is submerged will also effect your volume determination, so ensure that only the very tip of the rod is in the water. *Note*: If the solid sample is too large to go into the graduated cylinder, you can fill a suitable container, such as a large beaker, completely to the top. Be careful to catch ALL displaced water in the graduated cylinder as you carefully submerge the sample in the beaker, because any drops of water lost will cause a large error.
3. Calculate the density of the unknown solid using Eq. 27.1. Record the density in the data table. show your calculations on a separate sheet of paper. Use the appropriate number of significant digits in all calculations.

Data Table

Procedure 1

Mass of flask + stopper (also used in Procedure 2 and 3)	_____	g
Temperature of water	_____	°C
Density of water from Table 27.1	_____	gm/cm³
Mass of flask + stopper + water	_____	g
Mass of water in flask	_____	g
Calculated volume of flask	_____	cm³

Procedure 2

Mass of flask + stopper + unknown liquid	_____	g
Calculated mass of unknown liquid	_____	g
Volume of unknown liquid	_____	cm³
Calculated density of unknown liquid	_____	g/cm³

Procedure 3

Mass of flask + stopper + unknown solid	_____	g
Calculated mass of unknown solid	_____	g
Mass of flask + stopper + unknown solid + water	_____	g
Volume of unknown solid	_____	cm³
Calculated density of unknown solid	_____	g/cm³

Procedure 4

Mass of unknown solid (found on balance)	_____	g
Volume of water in graduated cylinder	_____	cm³
Volume of water + solid in graduated cylinder	_____	cm³
Volume of solid (found by displaced-water method)	_____	cm³
Density of unknown solid	_____	g/cm³

PROCEDURE 5

The purpose of this procedure is to determine the density of a liquid or solid by direct measurement of volume and mass. *Note:* For this procedure you will have to design your own data table.

1. Find the mass of the unknown material using a balance. If the material is a solid, the mass may be found by weighing the sample directly. If it is a liquid, you will have to weigh a suitable container when it is empty and then again when it contains the unknown liquid sample. The difference in mass readings will give the mass of the liquid sample. Record this mass in your data table.
2. Measure the volume of the unknown sample. If it is a regularly shaped solid, measure the appropriate dimensions of the object and then calculate its volume using the equations found in Table 27.2. If the material is a liquid, carefully pour it into a clean graduated cylinder and read its volume on the scale. Record this mass in your data table.
3. Calculate the density of your sample using Eq. 27.1 and record this value in your own data table.
4. Repeat this procedure for several liquid and/or solid samples and record all data and calculations in your data tables.

QUESTIONS

1. Define density and give an example, using SI units.

2. Determine the density of pure, air-free water at atmospheric pressure and having a temperature of (a) 2°C, (b) 22°C.

3. State the maximum density of water and give the temperature at which maximum density occurs.

4. Determine the mass of a solid that has a volume of 6.39 cm^3 and a density of 5.4 g/cm^3. Be sure that your answer contains the correct number of significant digits.

5. Determine the volume of a liquid that has a mass of 24.6 g and a density of 0.82 g/cm^3. Be sure the answer has the correct number of significant digits.

6. Explain how the volume of the human body can be determined experimentally.

Oxygen

INTRODUCTION

The element we now call oxygen was discovered in 1774 by Joseph Priestley (1733–1804), an English minister, who obtained the gas from mercuric oxide. The mercuric oxide was obtained by heating mercury with the "burning glass" (magnifying lens). The gas was distinguished as an element and named oxygen by A. L. Lavoisier.

Oxygen is the most abundant element in Earth's crust. The atmosphere contains 23% oxygen by weight, and water contains almost 89% oxygen by weight. Oxygen has an atomic number of 8, and because oxygen combines readily with many elements it was chosen as the standard to which the weights of other elements were compared. In 1961 oxygen was dropped as a standard, and the most abundant isotope of carbon was taken as the new standard. The standard value assigned to this isotope was exactly 12 atomic mass units.

LEARNING OBJECTIVES

After completing this experiment, you should be able to do the following:

1. Prepare a volume of oxygen from the compound potassium chlorate.
2. State several physical and chemical properties of oxygen.

APPARATUS

A mixture of about 80% potassium chlorate ($KClO_3$) and 20% manganese dioxide (MnO_2), one Pyrex test tube, ring stand and clamp, Bunsen burner, gas lighter, one-hole rubber stopper, delivery tube (rubber hose), pneumatic trough, four collecting bottles (500 mL), large shallow tray to catch water overflow from pneumatic trough, wood splints, sulfur (pinch), deflagrating spoon, iron picture wire (15 cm in length), four glass plates.

PROCEDURE

Fill the pneumatic trough with water. Completely fill four bottles with water, cover them with glass plates, invert them, submerge their necks in the water in the pneumatic trough and remove the glass covers. Allow no air bubbles. *Be sure that the pneumatic trough is set in the shallow tray provided for your table.*

You will need a mixture of about 80 percent potassium chlorate ($KClO_3$) and 20 percent manganese dioxide (MnO_2). Your instructor may provide these compounds already pre-mixed in the proper proportions. If you prepare your own mixture, observe the following warning.

Caution: Explosions can occur when combustibles such as carbon, sulfur, or rubber come into contact with fused potassium chlorate. Heat the manganese dioxide well before using, and inspect the potassium chlorate for any foreign materials.

Place 2 to 4 g of the $KClO_3$-MnO_2 mixture in a Pyrex test tube. Insert the stopper and delivery tube, arranging the assembly as shown in Fig. 28.1.

Figure 28.1

Adjust the Bunsen burner to a low flame and place under test tube so as to heat the mixture *slowly, heating the top part of the mixture first.* Do not collect gas in the bottles until you see the reaction beginning in the black mixture. **Important!** Heat so that the *oxygen bubbles through the water steadily but not too fast.* Remove the flame for a few moments if it bubbles too vigorously. When the water has been completely displaced by the gas, set the bottle aside and start filling the next.

When all the bottles are completely filled with gas, *pull the delivery tube from the trough* and then remove the bottles from the trough as follows. Raise the bottle enough to slip a glass plate over the mouth under water; hold the glass plate tightly against the mouth of the bottle; lift the bottle out and set it upright on the table. *Keep the bottle covered* until the oxygen is needed.

What has happened is this: The heat of the flame decomposes the potassium chlorate ($KClO_3$) producing potassium chloride (KCl) and oxygen gas (O_2), thus:

$$2 \, KClO_3 \rightarrow 2 \, KCl + 3 \, O_2$$

The manganese dioxide has acted as a catalytic agent; that is, it has undergone no chemical change itself but it has promoted a chemical reaction, in this case, the decomposition of the potassium chlorate.

Tests with oxygen

1. Ignite a wood splint by means of your Bunsen burner. When the splint has burned for a moment, blow out the flame, leaving the splint still glowing. Plunge the glowing splint into a bottle of oxygen. Perform this procedure at least one more time. Describe what happens.

2. Heat a **small** pinch of sulfur in a deflagrating spoon until the sulfur begins to burn. Observe the color and size of the flame. Now, lower the spoon of burning sulfur into a bottle of oxygen. Compare the size and color of the flame in air and in oxygen. (Put out the flame by submerging the spoon and sulfur in water as soon as the experiment is finished.)

3. Heat the tip of the iron picture wire directly above the tip of the blue inner cone of the Bunsen burner flame until it becomes red-hot. **Quickly** plunge it into a bottle of oxygen. Record your observations.

4. Take a pinch of sulfur on a dry glass plate. Again heat the tip of the iron picture wire red-hot and at once dip it into the pinch of sulfur. While the sulfur is still burning quickly thrust this end of the wire into a jar of oxygen. What do you observe? A reaction should take place. If you fail to see any reaction, notify the instructor.

QUESTIONS

1. What do you think holds the water in the bottles that, after being filled, are inverted in the pneumatic trough?

2. From the results of Tests 1 and 2 in this experiment, what statement could be made about the effect of oxygen on the process of combustion?

3. When sulfur is oxidized (combined with oxygen), of what chemical elements is the product composed?

4. In Test 4, the sulfur does not become part of the final product. Comparing Tests 3 and 4 of this experiment, what was the purpose of the sulfur in Test 4?

5. In Test 4, what happened to the iron wire? Of what chemical elements was the product of the reaction composed?

Percentage of Oxygen in Potassium Chlorate

INTRODUCTION

The percentage composition of a compound can be calculated if the formula for the compound is known. For example, the percentage composition of sulfuric acid (H_2SO_4) is calculated as follows:

1. Write the chemical formula.

$$H_2SO_4$$

2. Look up relative weights and place over symbols.

$$\begin{array}{ccc} 1 & 32 & 16 \\ H_2 & S & O_4 \end{array}$$

3. Determine total relative weight of each element present by multiplying relative weight by subscript.

$$\begin{array}{ccc} H_2 & S & O_4 \\ 2 & 32 & 64 \end{array}$$

4. Add total relative weights and indicate each element as a fraction of the total.

$$\begin{array}{ccc} H_2 & S & O_4 \end{array}$$
$$2/98 + 32/98 + 64/98 = 98/98$$

5. Express each fractional part as a percentage by multiplying the fraction by 100.

$$2/98 \times 100/1 = 2.04\%$$
$$32/98 \times 100/1 = 32.65\%$$
$$64/98 \times 100/1 = 65.31\%$$

LEARNING OBJECTIVES

After completing this experiment, you should be able to do the following:

1. Determine the percentage composition of a compound.
2. Determine the theoretical and experimental values of oxygen in the compound potassium chlorate.

APPARATUS

Potassium chlorate, manganese dioxide, Pyrex test tube, ring stand, test-tube clamp, Bunsen burner, gas lighter, beam balance.

PROCEDURE

Weigh a clean, dry test tube with a few grains of manganese dioxide (MnO_2) in it.* Record weight in Data Table 29.1. Place in the tube about one gram of potassium chlorate ($KClO_3$) and weigh again. Put the test tube in the clamp provided on the stand, and cover its mouth with a crucible cover. Using a clean, nonluminous flame, heat the test tube, including the sides, gently for a few minutes, then more strongly. Avoid the occurrence of white fumes in the tube. Continue the heating until the reaction stops. Cool the tube and weigh it again. Resume heating for a few minutes; again cool, and reweigh. If the two weights are close in value, the process is complete. If not, heat until a constant weight is obtained on two successive weighings. The difference between the weight before heating and the constant weight after heating represents the weight of oxygen driven off.

Calculate from your data the percentage of oxygen in potassium chlorate.

Data Table 29.1

Weight of test tube, MnO_2, and $KClO_3$ before heating	_____	g
Weight of test tube and MnO_2	_____	g
Weight of test tube and residue after heating	_____	g
COMPUTATIONS Weight of $KClO_3$	_____	g
Weight of oxygen driven off	_____	g
Percentage of oxygen in $KClO_3$	_____	%

From the formula $KClO_3$ and the known atomic weights of its elements, calculate the theoretical percentage of oxygen in potassium chlorate. Compare this with the value obtained in your experiment by computing percent error; the percent error should be small.

Theoretical percentage of oxygen in $KClO_3$ .... _____ %

Percent error .... _____ %

QUESTIONS

1. What is the purpose of the manganese dioxide in this experiment?

*The instructor will dispense the MnO_2.

2. What is the theoretical percentage of Cl in $KClO_3$? Show your calculations.

3. What is the theoretical percentage of K in $KClO_3$? Show your calculations.

4. Where is the greatest possible source of error in this experiment? Why?

5. Balance the chemical equation:

$$\underline{}\ KClO_3 \rightarrow \underline{}\ KCl + \underline{}\ O_2$$

Experiment 30

Percentage of Oxygen and Nitrogen in the Air

INTRODUCTION

The atmosphere is composed almost entirely of chemically uncombined nitrogen and oxygen. There are small amounts of argon, krypton, neon, water vapor, carbon dioxide, and dust. During chemical oxidation processes in which metal oxides are formed in the air, oxygen is used up, leaving nitrogen as the chief constituent of the remaining air. Such a process, carried out carefully, can be used to determine the percentage of oxygen and nitrogen in the air.

The general procedure is to pass a measured quantity of air *slowly* over a *red-hot metal*, and after the oxygen is removed in the formation of an oxide, measure the quantity of nitrogen gas that is left. Copper metal serves the purpose very well. Hot copper combines with the oxygen of the air to form a black copper oxide:

$$2\,Cu + O_2 \rightarrow 2\,CuO$$

LEARNING OBJECTIVE

After completing this experiment, you should be able to determine the percentage of oxygen and nitrogen in the air.

APPARATUS

Combustion tube, Bunsen burner, wing tip, thistle tube, one 500-mL bottle, one 300-mL bottle, pneumatic trough, large shallow tray, two ring stands, two burette clamps, two one-hole rubber stoppers, one two-hole rubber stopper, rubber hose, pinch clamp, copper turnings, graduated cylinder.

PROCEDURE

1. *Preparation*: Fill the pneumatic trough nearly full of water. Fill a bottle to the brim with water and place it in the trough. Be certain that there is no air in the bottle before the experiment is started.

   Assemble the apparatus as shown in Fig. 30.1. *Be sure that the delivery tube is not under the collecting bottle B_2 when the experiment begins.*

Figure 30.1

   Without applying any heat to the combustion tube, slowly pour about 100 mL of water into the thistle tube. As the water drips into the bottle B_1, air will be forced very slowly through the combustion tube and will bubble from the end of the delivery tube. If it does not, check for leaks. Tighten carefully all clamps, connections, and stoppers. Adjust the bubbling rate by means of the pinch clamp to about two bubbles per second.

   When the water has completely gone through the thistle tube, put the flame under the combustion tube and begin heating the tube. A few bubbles due to expansion of air in the combustion tube will come out of the end of the delivery tube. When these have ceased to appear, put the end of the delivery tube under bottle B_2. Accurately measure out in a graduated cylinder 300 mL of water. Pour this into the thistle tube a little at a time (do not let the thistle tube top become empty at any time) until all 300 mL are in the bottle. Note that this second batch of water will force 300 mL of air from the bottle B_1 through the combustion tube. The nitrogen residue from this 300 mL of air will be collected in bottle B_2.

   When all the water has drained from the thistle tube, and the bubbles cease appearing in bottle B_2, remove the delivery tube from the trough and the burner from under the combustion tube. Slip a cover glass carefully over the mouth of bottle B_2, remove the bottle from the trough and set it upright on the table. Empty the water from B_1 and make a second trial.

2. To obtain the percentage of nitrogen in the air, fill a graduated cylinder with water to its highest reading. Pour some of this water into bottle B_2, which holds the nitrogen gas. *Fill B_2 up to the brim.* Record the number of mL of water required. This is the volume of nitrogen in 300 mL of air.

Data Table 30.1

	First Trial	Second Trial
Number of mL of nitrogen		
Number of mL of air forced through		
COMPUTATIONS Percentage of nitrogen in air		
Percentage of oxygen in air		

QUESTIONS

1. Why was the air forced slowly through the combustion tube?

2. If part of the 300 mL of water were lost by an overflow of the thistle tube, how would the results of the experiment be affected?

3. If an air leak developed at the right end of the combustion tube, how would the results of the experiment be affected?

An Exothermic Chemical Reaction

INTRODUCTION

A chemical reaction produces a chemical change that yields one or more new substances. During a chemical reaction, energy may be absorbed or released. When energy is absorbed, the reaction is called an **endothermic chemical reaction**. When energy is released the reaction is called an **exothermic chemical reaction**. The energy released may be in the form of heat, light, or electricity. An example of heat released is the burning of a match.

Heat is a form of energy (energy in transit) that cannot be measured directly. In this experiment, an indirect measurement of the amount of heat released during the reaction of hydrochloric acid (HCl) and sodium hydroxide (NaOH, a base) will be obtained by measuring the temperature change that occurs when the two substances are mixed. The reaction is represented empirically by the following equation:

$$HCl + NaOH \rightarrow NaCl + H_2O$$

The energy released during a chemical reaction is directly proportional to the concentration of the reacting substances. Another way of stating this is: *the energy released is directly proportional to the number of molecules taking part in the chemical reaction*. The amount of heat released during a chemical reaction is also a function of the nature of the substances taking part in the reaction.

A one-molar (1.0-M) solution is one mole of the substance per liter of solution. One mole of anything contains Avogadro's number (6.023×10^{23}) of particles. A 2.0-molar solution contains two moles per liter of solution and contains 12.046×10^{23} particles. In this experiment 2.0-M solutions of NaOH and HCl will be used. The 2.0-M solutions will be referred to as the "standard solutions."

In this experiment, everything is held constant except the strength (number of molecules per cubic centimeter) of the hydrochloric acid and the sodium hydroxide. As the strength of these two solutions is varied by diluting with water, the temperature changes will be measured and recorded. The temperature changes during the mixing of the two solutions of different concentrations will give an indication of how much heat is released during the chemical reaction. This is true because the heat released is proportional to the change in the temperature.

LEARNING OBJECTIVES

After completing this experiment, you should be able to do the following:

1. Define an exothermic chemical reaction and give an example.
2. State the relationship between temperature changes and the concentration of the reacting solutions that occur in a chemical reaction.
3. State the relationship between the energy released in a chemical reaction and the concentration of the reacting solutions.

APPARATUS

Calorimeter, thermometer, two graduated cylinders 50 ml, one beaker 25 mL, distilled water, 2.0-M sodium hydroxide solution, 2.0-M hydrochloric acid solution.

PROCEDURE

The purpose of the procedure is to collect data to determine the relationship between the concentration of the reacting solutions (sodium hydroxide and hydrochloric acid) and the amount of heat released during the chemical reaction that occurs when the two solutions are mixed.

The volumes of the two solutions will be kept constant at 20 mL, and the concentration of the two solutions varied. The original temperature of the sodium hydroxide and the hydrochloric acid will be measured and held constant, preferably at room temperature. When the two solutions are mixed, the temperature change will be measured and recorded.

Step 1   Measure 20 mL of the standard solution of sodium hydroxide* and place it in the calorimeter. Measure and record the temperature of the 20-mL solution; it should be at room temperature. This is the original temperature T_o.

Step 2   Measure 20 mL of the standard solution of hydrochloric acid* and transfer it to the 25-mL beaker. Measure and record the temperature of the HCl; it should also be at room temperature.

Step 3   Transfer and mix the HCl solution with the NaOH in the calorimeter. Stir the mixture with the thermometer as the HCl is added. Measure and record the highest temperature of the mixture. This will be recorded as T_h in the data table. The highest temperature reading will occur almost immediately upon mixing the two solutions. The temperature reading will be slightly lower than the true temperature, because (1) some heat is lost to the surroundings and (2) it takes time for the mercury to rise in the thermometer.

Step 4   Reduce the concentration of the NaOH and HCl solutions. Place 15 mL of the standard solution of sodium hydroxide in the graduated cylinder. Add 5 mL of distilled water and transfer the 20 mL (75% solution) to the calorimeter. Place 15 mL of the standard solution of HCl in the other graduated cylinder. Add 5 mL of distilled water and transfer the 20 mL (75% solution) to the 25-mL beaker.

Step 5   Repeat Step 3.

Step 6   Reduce the concentration of the NaOH and HCl solutions. Place 10 mL of the standard NaOH solution in the graduated cylinder. Add 10 mL of distilled water and transfer the 20 mL (50% solution) to the calorimeter. Place 10 mL of the standard HCl solution in the other graduated cylinder. Add 10 mL of distilled water and transfer the 20 mL (50% solution) to the 25-mL beaker.

Step 7   Repeat Step 3.

Step 8   Reduce the concentration of the NaOH and HCl solutions. Place 5 mL of the standard NaOH solution in the graduated cylinder. Add 15 mL of distilled water and transfer the 20 mL (25% solution) to the calorimeter. Place 5 mL of the standard HCl solution in the other graduated cylinder. Add 15 mL of distilled water and transfer the 20 mL (25% solution) to the 25-mL beaker.

Step 9   Repeat Step 3.

Step 10   Reduce the concentration of the NaOH and HCl solutions. Place 2.5 mL of the standard NaOH solution in the graduated cylinder. Add 17.5 mL of distilled water and transfer the 20 mL

---

*Use separate graduated cylinders to measure the volume of sodium hydroxide and hydrochloric acid.

(12.5% solution) to the calorimeter. Place 2.5 mL of the standard HCl solution in the other graduated cylinder. Add 17.5 mL of distilled water and transfer the 20 mL (12.5% solution) to the 25-mL beaker.

Step 11 Repeat Step 3.

Step 12 Analyze the data by means of a graph. Plot the temperature changes on the y axis and the percent concentration on the x axis. Refer to Experiment 1 for information on how to correctly plot and label a graph. Should the intersection of the x and y axes be a point on your graph? (*Hint:* What would be the temperature change when the percent concentration is zero?)

Data Table 31.1

Concentration of Solutions	T_o	T_h	ΔT
100%			
75%			
50%			
25%			
12.5%			

T_o = original temperature of the NaOH and HCl solutions
T_h = highest temperature of the mixed solution
ΔT = Change in temperature ($T_h - T_o$)

QUESTIONS

1. What does the graph indicate concerning the relationship between percent concentration of solution and the changes in temperature?

2. Explain the relationship between the heat released by the chemical reaction and the change in temperature.

3. Explain the slope of your graph.

4. How would the slope of your graph change if the volume of the solutions were doubled? Explain your answer.

5. Assume that the chemical reaction between the NaOH and the HCl is completed. (a) With the 100% standard solution, how many sodium ions react with how many chlorine ions? (b) With the 25% solution, how many water molecules were formed?

Experiment 32

Avogadro's Number

INTRODUCTION

The elements that make up our environment vary greatly in their physical and chemical properties and in the types of compounds they form. Elements with similar properties can be classified and placed in groups or arranged in order relative to some physical or chemical property. The periodic table is one method used to group the elements. This grouping is based upon the **periodic law**, which states that the physical and chemical properties of the elements are periodic functions of their atomic number. That is, the properties of elements appear in a certain order, and elements with similar properties occur at particular intervals when arranged in order of increasing atomic number.

The periodic table also gives the relative weight of each element. The relative weight is given with respect to the carbon-12 atom, which was established as the reference standard by the International Union of Pure and Applied Chemistry in 1961. The relative weights are proportional to the actual weights of the atoms.

The relative weight of a molecule can be obtained by adding the relative atomic weights of the atoms that combine to form the molecule.

In work requiring weights of elements, it is convenient to use grams to express relative weight. The concept **gram-atomic weight** is defined as the amount of the element, expressed in grams, that is numerically equal to the relative atomic weight. For example, the gram-atomic weight of hydrogen is 1.0080 g and of oxygen, 15.9994 g. Similarly, the **gram-molecular weight** is the mass of a molecular substance in grams equal to its molecular weight. For example, the gram-molecular weight of water (H_2O) is 18.0154.

One gram-atomic weight of any element contains the same number of atoms. This number is called **one mole of atoms** and has the value of 6.022×10^{23}. One mole of atoms is defined as the number of atoms in exactly 12 grams of carbon-12.

The number 6.022×10^{23} is known as **Avogadro's number**. Amedeo Avogadro (1776–1856), an Italian physicist, announced in 1811 a hypothesis to explain the behavior of gas, which stated that under conditions of equal pressure and temperature, equal volumes of gases contain equal numbers of molecules.

Several methods can be used to determine Avogadro's number (symbol, N). In this experiment, the method of electrolysis will be used.

A typical example of electrolysis is shown in Fig. 32.1, which utilizes a solution of copper sulfate containing Cu^{++} ions and SO_4^{--} ions. When the switch S is closed, an electric field is placed between the copper electrodes; the Cu^{++} ions are drawn toward the negatively charged cathode, and the SO_4^{--} ions are drawn toward the positively charged anode. If the action is allowed to continue for a period of time, we should find that a mass m of metallic copper has been deposited on the cathode and an equal mass m of metallic copper has been removed from the anode. The concentration of the solution remains the same.

Figure 32.1

LEARNING OBJECTIVES

After completing this experiment, you should be able to do the following:

1. Define gram-atomic weight, gram-molecular weight, mole, and Avogadro's number.
2. Determine experimentally the numerical value of Avogadro's number.

APPARATUS

Coulometer, good milliammeter (range 0–500 or 0–1000 mA), copper electrodes, 40-ohm rheostat, copper sulfate solution containing 100 g copper sulfate, 18 mL concentrated H_2SO_4, 34 mL ethyl alcohol in 1000 mL distilled water (*Note*: filter through glass wool), one SPST switch, stirring motor, glass stirrer.

PROCEDURE

1. Connect the circuit as shown in Fig. 32.1.
2. If copper sulfate solution is not in the coulometer, obtain it from the instructor and fill the glass container.
3. Close switch S and adjust the rheostat R until the milliammeter reads 400 to 800 milliamps; then open the switch. The circuit is now adjusted for operation.
4. Remove the anode and cathode from the coulometer, dry them with paper toweling, polish them with steel wool, wipe them clean with soft paper and weigh each on a balance to ±0.01 gram.
5. Position the anode and cathode in the coulometer, close the switch S, start the timer (your watch will do) and the stirring motor, adjust the rheostat for 400 to 800 milliamperes, and *keep the current constant* for one hour.
6. At the end of one hour, open the switch S and quickly remove the anode and cathode, wash them in running water, rinse them with acetone, and air dry.
7. Weigh the anode and cathode to ±0.01 gram.
8. Record data in Data Table 32.1.

Data Table 32.1

1.	Current		A
2.	Time		s
3.	Loss of mass of anode (original mass – final mass)		g
4.	Gain of mass of cathode (final mass – original mass)		g
5.	Difference between loss and gain of mass		g

CALCULATIONS

1. Determine the amount of flow of charge q. (*Note: $q = It$.*)

_____ coulombs

2. Determine the number of electrons n_e flowing through the solution, where

$$n_e = \frac{q}{1.6 \times 10^{-19} \text{ coulomb}}$$

_____ electrons

3. Determine the number of copper atoms that were given up by the anode. The charge carried by each atom is +2.

_____ copper atoms

4. Determine the number of copper atoms N in 1 mole of copper. There are 63.54 g of copper in 1 mole.

$N =$ _____

5. Determine your percent error.

_____ %

QUESTIONS

1. Define one mole and give an example.

2. State the numerical value of Avogadro's number.

3. How many sodium ions are there in one mole of sodium chloride (NaCl)? How many chloride ions?

4. How many molecules are there in 47.8 grams of $CuSO_4$?

5. Where is the greatest possible source of error in this experiment? Why?

Molecular Structure

INTRODUCTION

A covalent bond is formed between two atoms when they share electrons. Covalently bonded molecules are formed by these atoms because each atom tends toward a stable configuration, and the greatest stability is achieved when the outer shell of electrons is closed, or has its full complement of electrons orbiting the nucleus. In the case of hydrogen, one electron orbits the nucleus of the atom, but two electrons are required to complete the shell and render stability. Two hydrogen atoms tend to share electrons to form a molecule of hydrogen gas.

$$H\cdot\cdot H$$

Dots between atoms indicate that the electrons are shared. Each hydrogen atom in a molecule has two electrons going around it.

If we concentrate on the lighter elements, ignoring the d and f electrons, we find that it takes eight electrons to make a closed shell (except for hydrogen and helium, in which two electrons form a closed shell). In the electron dot notation, we draw a dot for every electron outside closed shells. Thus, carbon, nitrogen, and chlorine, which have four, five, and seven electrons outside closed shells, are represented as:

$$\cdot \overset{\cdot}{\underset{\cdot}{C}} \cdot \qquad :\overset{\cdot}{\underset{\cdot}{N}} \cdot \qquad \cdot \overset{\cdot\cdot}{\underset{\cdot\cdot}{Cl}} :$$

For a carbon atom to make a stable molecule, it must obtain four electrons from other atoms. Nitrogen needs three, and chlorine only one. Simple arrangements of molecules of carbon, nitrogen, and chlorine each united with hydrogen are CH_4, NH_3, and HCl. They are expressed in the electron dot notation as:

Note that each atom in these compounds has a closed shell of electrons.

Because it has four unpaired electrons, carbon forms an unusually large number of different compounds. Carbon compounds are called **organic compounds**. Many of the organic compounds are quite complicated. To portray them, a line is frequently used to indicate two electrons. For example, methane, CH_4, and ethane, C_2H_6, are drawn as in the accompanying structural formulas.

Methane   Ethane

Double and triple bonds are formed when four or six electrons are shared among atoms. Ethylene, C_2H_4, and acetylene, C_2H_2, are examples of simple molecules with double or triple bonds. Their structural formulas are shown here.

Ethylene   Acetylene

Note that each carbon and hydrogen has a filled shell of electrons, as it must.

Carbon compounds such as cyclopropane, C_3H_6, can also be formed in ring structures.

Cyclopropane

More complicated compounds of carbon have oxygen, nitrogen, the halogens, and other elements combined in complicated configurations.

When two compounds have exactly the same chemical constituents but different structures, they are called isomers. Two isomers of C_3H_6 are cyclopropane and propene, whose structures are shown.

Propene

In the study of isomers, it is necessary to understand that two atoms joined by a single bond are each free to rotate independently about the axis formed by the bond.

Free rotation about C–C bond

With a double bond, there is restricted rotation.

No rotation about C=C bond

In order to study molecular structure effectively, we must work with three-dimensional molecules. Hence, we need models in three dimensions. For example, CH_4 in three dimensions has the hydrogens at the corners of a tetrahedron with the carbon in the middle.

Two-dimensional representation   Three-dimensional representation

LEARNING OBJECTIVES

After completing this experiment, you should be able to do the following:

1. Define the terms covalent bond and isomers.

2. Construct models of molecules using a molecular models kit.

APPARATUS

A molecular models kit consisting of beads with holes inserted, spring connectors, and a protractor.

Table 33.1 gives the color code information on the molecular models kit. Each different-colored bead can be used to represent a different atom. Each hole in a bead represents an unpaired electron.

Each spring represents a covalent bond of two electrons. A single bond is represented by one spring, a double bond by two springs. Use the short spring for C–H bonds and the longer springs for other bonds.

The models should be assembled by insertion of the springs with a partial clockwise turn. In order to avoid damage to the springs, the models should be dismantled by withdrawing the springs with the same clockwise turn.

Table 33.1   Color Code Information, Molecular Models Kit

Color of bead	Number of holes	Element represented
Black	4 holes	Carbon
White	1 hole	Hydrogen
Red	2 holes	Oxygen
Green	1 hole	A halogen (fluorine, chlorine, bromine, or iodine)
Orange or yellow	4 holes	Anything

PROCEDURE

1. Make a model of CH_4. Measure the angle between the two bonds with your protractor. Record it.
   Bond angle ....   _____

2. Make a model of CH_3Cl. Are there 2 isomers with configurations as drawn below? Or are these the same?

3. Make models of CH_2ClBr. Are there two isomers as shown below? Or are these the same?

4. Are the models in Steps 2 and 3 planar?..   _____
   -

5. Make a model of C_2H_5Cl.

   Is this molecule planar?...   _____

   How many isomers are there?...   _____

6. Make a model of C_2H_3Cl.

   Is this molecule planar?...   _____

   How many isomers are there?...   _____

7. Make as many isomers of C_3H_8 as you can. One of them is

   How many are there? ....   _____

8. Make models for all isomers of C_3H_6. How many are there? In the space below, draw structural formulas for all the isomers. (If you need more space, use the blank page at the end of the chapter.) .... _____

9. Make models for all the isomers of C_3H_5Cl. How many are there? In the space below, draw structural formulas for all the isomers. (If you need more space, use the blank page at the end of the chapter.) .... _____

10. Make models for all the isomers of CHBrClF. How many are there? One is

11. Make models for all the isomers of C_4H_{10}. How many are there? .... _____

12. Make models for all the isomers of C_4H_9Cl. How many are there? .... _____

13. Make models for all the isomers of C_5H_{12}. How many are there? .... _____

14. Make a model of ethyl alcohol. Its structure is

15. Make a benzene ring model. Its structure is

Have the instructor inspect it and check here .... _____

16. Make a long-chain molecule of $C_{10}H_{22}$. Could you make an indefinitely
long chain? _____

17. Make a model of nicotine. Its structure is

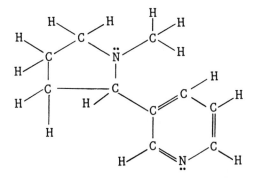

Have the instructor inspect your model and check here.................................. _____

243

Solutions and Solubility

INTRODUCTION

A **mixture** is a sample of matter composed of two or more substances in varying amounts that are not chemically combined. Mixtures can be homogeneous or heterogeneous. A **solution** is a homogeneous mixture of two or more substances. The substance present in greatest quantity in a solution is called the **solvent**. The substance dissolved is the **solute**. Solutions in which water is the solvent are called **aqueous** solutions. A solution in which the solute is present in only a small amount is called a **dilute** solution. If the solute is present in a large amount, the solution is a **concentrated solution**. When the maximum amount of solute possible is dissolved in the solvent, the solution is called a **saturated** solution. The concentration is frequently expressed in terms of the number of grams of solute dissolved in a given quantity of solvent. Example: a solution of salt in water might contain 5 g of salt for each 100 g of water.

The **solubility** of a given solute is the amount of solute that will dissolve in a specified volume of solvent (at a given temperature) to produce a saturated solution. The solubility depends on the temperature of the solution. If the temperature increases, the solubility of the solute will almost always increase.

LEARNING OBJECTIVES

After completing this experiment, you should be able to do the following:

1. Define mixture, solvent, solute, solution, and solubility.
2. Determine the concentration of a salt in water solution.
3. Determine the solubility of a substance as a function of temperature.

APPARATUS

Salt in water solution of unknown concentration, salt, distilled water, evaporating dish, glass beaker, ring stand with holder to support beaker, thermometer, Bunsen burner, lighter, and beam balance.

PROCEDURE

1. The concentration of a solution can be determined by separating the solute from the solvent. Find the weight of each, then calculate the ratio of grams of solute to 100 g of solvent.
   (a) Weigh a clean, dry evaporating dish.
   (b) Obtain from the instructor about 4 cm³ of a salt solution of unknown concentration. Pour the solution into the evaporating dish.
   (c) Weigh the dish plus solution. Record in Data Table 34.1.
   (d) Place the evaporating dish on a beaker of gently boiling water and let the water evaporate from the salt.
   (e) When the water has completely evaporated from the dish, remove the dish from the beaker, cool, then remove any moisture from the bottom of the dish. Weigh the dish and residue. Record in the data table.
   (f) Determine from your data the weight of the solute and the weight of the solvent in your sample.
   (g) Determine the concentration as grams of solute per 100 g solvent.

2. The solubility of a substance as a function of temperature can best be done in a limited time period by using one substance and having each group of students at a laboratory table obtain the solubility at a specific temperature and share the results. The following temperatures are suggested: room temperature which will usually be about 22°C, 30°C, 40°C, 50°C, 60°C, 70°C. If there are more or less than six groups, duplicate a temperature or let a group do two temperatures. Use two sets of apparatus to save time if a group is to do two temperatures.
   (a) Place about 20 mL distilled water in a clean beaker and bring to the correct temperature. Keep the temperature constant and add small amounts of salt while stirring continuously. Add salt until it fails to dissolve in the water.
   (b) Keep the temperature constant and allow the excess salt to settle to the bottom of the beaker. Weigh a clean, dry evaporating dish and pour about 4 mL of the solution, still at the constant temperature, into the dish.
   (c) Weigh the dish plus the solution. Record in Data Table 34.1.
   (d) Place the evaporating dish on a beaker of gently boiling water and let the water evaporate from the salt.
   (e) When the water has completely evaporated from the dish, remove the dish from the beaker, cool, then remove any moisture from the bottom of the dish. Weigh the dish and residue. Record in the data table.
   (f) Determine from your data the weight of the solute and the weight of the solvent in your sample.
   (g) Determine the concentration as grams of solute per 100 g of solvent.
   (h) Plot a graph of solubility as a function of temperature. Plot solubility on the y axis and temperature on the x axis.

Data Table 34.1

Part 1

Weight of clean, dry evaporating dish	_____ g
Weight of evaporating dish plus solution	_____ g
Weight of evaporating dish plus dry residue	_____ g
Weight of solvent (computed)	_____ g
Weight of solute (computed)	_____ g
Grams of solute per 100 g of solvent (computed)	_____ g

Part 2

Temperature	Room _____ °C	30°C	40°C	50°C	60°C	70°C
Weight of clean, dry dish						
Weight of dish plus solution						
Weight of dish plus dry residue						

COMPUTATIONS

Temperature	Room _____ °C	30°C	40°C	50°C	60°C	70°C
Weight of solute						
Weight of solvent						
Grams of solute per 100 g of solvent						

QUESTIONS

1. Define the term solution. Give an example.

2. What is a saturated solution?

3. Distinguish between solvent and solute.

4. Define solubility. Give an example.

5. Is solubility a function of temperature? Explain.

Pressure-Volume Relationship of Gases

INTRODUCTION

A gas is a substance that has no definite shape or volume. If we investigate a sample of any gas, we discover that it fills the entire volume of the container it occupies. Close examination reveals the gas molecules to be in rapid motion, undergoing collisions with one another and with the walls of the container.

The behavior of gases is understandable if we contemplate the ideal gas. Such a gas would be composed of molecules that are small in size compared with the total volume of a container and where molecules have no attraction for one another. This description fits most of the gases found in our environment.

The physical states used to describe the gases are volume, pressure, and temperature. The relationships of these states and the laws governing such relationships were formulated by Robert Boyle in the 16th century, and Jacques Charles and Joseph Gay-Lussac in the 17th century.

The absolute temperature of a gas can be defined as a measure of the average kinetic energy of the gas molecules:

$$T \propto \frac{mv^2}{2}$$

where T = absolute temperature,

m = mass,

v = velocity.

Pressure can be defined as the force per unit area:

$$P = \frac{F}{A}$$

The **general gas law**, which is a combination of Boyle's and Charles's laws, deals with the expansion and compression of a gas at different temperatures. The law states that the pressure P of a given amount of gas is proportional to the absolute temperature T, and inversely proportional to the volume V, or

$$\frac{PV}{T} = k$$

where the constant k has a different value for different gases. When samples of the same gas are taken that have different total masses m, we find experimentally that

$$\frac{PV}{T} \propto m$$

When equal masses of gases are taken that have different molecular weights M_w, we find that

$$\frac{PV}{T} \propto \frac{1}{M_w}$$

thus,

$$\frac{PV}{T} \propto \frac{m}{M_w}$$

But m/M_w is the number of moles n in the sample (constant throughout this experiment). Therefore,

$$\frac{PV}{T} \propto n$$

or

$$\frac{PV}{T} = Rn$$

where R is known as the universal gas constant.

The numerical value of the constant can be evaluated by considering a gas at standard temperature ($273°K$) and standard pressure (1 atmosphere = 14.7 lb/in^2 = 76 cm of mercury (Hg) = 1.013×10^5 newtons/meter^2). The measured volume of any gas at these conditions is approximately $22.4n$ L, where n is the number of moles, and $1\ L = 1 \times 10^{-3}\ m^3$. Thus we obtain

$$\frac{PV}{T} = \frac{P_0 V_0}{T_0}$$

where the zero subscript indicates standard conditions.

Substituting known values into the equation yields

$$\frac{PV}{T} = \frac{(1.013 \times 10^5 \text{ newtons/meter}^2) \times (22.4n) \times 10^{-3} \text{ meter}^3}{273°K}$$

or

$$\frac{PV}{T} = 8.31n \frac{\text{newton} \times \text{meter}}{°K}$$

so

If $n = 1$ mole $\qquad\qquad R = \dfrac{PV}{T} = 8.31 \dfrac{\text{newton} \times \text{meter}}{°K}$

LEARNING OBJECTIVES

After completing this experiment, you should be able to do the following:

1. Define the terms gas, pressure, and absolute temperature.
2. State in words the general gas law, and Boyle's law.
3. Determine experimentally the relationship that exists between pressure and volume of a gas when the temperature is held constant.

APPARATUS

In this experiment we use the Boyle's law apparatus (Fig. 35.1), or a substitute apparatus purchased from a distributor of scientific equipment.

Figure 35.1 *Apparatus for demonstrating Boyle's law.*

DISCUSSION

Fig. 35.1 is a drawing of the Boyle's law apparatus used in this experiment. The air pressure in the steel container is given by the pressure gauge and is practically the same as the pressure of the air in the glass tube. The difference between the two is due to the difference between the oil levels. Aside from this small difference, the pressure in the glass tube is the same as the pressure indicated by the pressure gauge. The meter scale is positioned to read the length of the air column in the glass tube. The volume V of the trapped air equals the cross-sectional area of the tube times the length of the air column. Because the cross-sectional area is constant, the volume is proportional to the length h of the air column. Thus, a method is available for measuring the volume and the pressure of the air in the glass tube.

When the valve is open, air is allowed to escape from the steel container; thus the pressure is lowered and the volume of air in the glass tube allowed to increase. If sufficient time (2 or 3 min) is allowed to pass before the pressure and volume are read, the temperature of the enclosed air will return to room temperature, and data concerning pressure versus volume can be obtained at constant temperature.

PROCEDURE

1. Inflate the Boyle's-law apparatus with air to a pressure of 90 lb/in^2 if this has not been done by the instructor.
2. Read the gauge pressure and the oil-level position and record all significant figures in Data Table 35.1.
3. Open the valve to the steel container and allow the pressure to decrease 5 to 10 lb/in^2. Allow 2 or 3 min for the air to return to room temperature and the oil in the glass tube to run from the side of the glass. Read and record the gauge pressure and the oil level.
4. Continue this process of decreasing the pressure a few pounds per square inch at intervals until all the air has been let out of the steel container and several readings of pressure and volume have been made and recorded in the data table.
5. Obtain the existing atmospheric pressure from the instructor.

Atmospheric pressure (P_{atm}) .... _____ cm Hg

The absolute or total pressure of the gas in the container is found by adding the gauge pressure to the atmospheric pressure:

$$P_{absolute} = P_{gauge} + P_{atmospheric}$$

Because 76 cm Hg equals 14.7 lb/in^2, gauge pressure in cm Hg can be calculated by multiplying the gauge pressure in lb/in^2 by 5.17, that is, 76 divided by 14.7.

Data Table 35.1

Gauge pressure, in lb/in^2	Gauge pressure, in cm Hg	Absolute pressure, in cm Hg	Length of air column in cm	Product of absolute pressure and length of air column

CALCULATIONS

1. Perform the calculations called for in the data table.
2. Plot a graph of absolute pressure versus $1/V$. The volume is indicated by the length of the air column. Plot the pressure on the y-axis.

QUESTIONS

1. Explain the curve you have plotted.

2. State the relationship between the volume and pressure of a gas, when the temperature is held constant. (Boyle's law)

3. State the relationship between the volume and temperature of a gas, when the pressure is held constant. (Charles's law)

4. The temperature of a gas is increased while the volume is held constant. Does the pressure decrease, increase, or remain the same? Explain.

5. A volume of gas is decreased by half while the temperature remains constant. How much does the gas pressure change?

Chemical Qualitative Analysis

INTRODUCTION

The systematic effort to determine what elements are present in a sample of unknown constituency is called **qualitative analysis**. Such elements can be determined by using the knowledge we have about their chemical characteristics. The procedure may be illustrated with a test for the metals **lead, silver,** and **mercury**, commonly known as **Group I**, with the common characteristic that their chlorides are largely insoluble in water. In a solution of their compounds, these elements can be detected by adding hydrogen chloride to the solution; the metals will be precipitated as chlorides and may be filtered out and thereby separated from any other metals present.

Lead chloride is very soluble in hot water and may be separated from the other precipitated chlorides by washing the mixture with hot water. The presence of silver can then be detected by adding ammonium hydroxide to the mixture. The silver ion will combine with ammonia to form a complex silver ammonium ion, whose chloride is soluble; the silver is thereby separated and put into solution as a complex ion. The mercury chloride that is left can be separated by filtration.

The presence of each of these elements in solution can be confirmed as follows: If potassium chromate is added to the hot water solution, a yellow precipitate will confirm the presence of lead. The presence of the silver in the filtrate obtained after the ammonium hydroxide is added may be confirmed by adding to the filtrate some nitric acid. This neutralizes the ammonium hydroxide and silver chloride is precipitated. The presence of mercury is confirmed immediately when ammonium hydroxide is first added to the mixture, since black, finely divided mercury is produced along with the ammonium-mercuric ion, both of which remain on the filter when the silver ammonium compound is filtered through in solution. A flow diagram summarizing the above instructions is provided as a part of these instructions.

LEARNING OBJECTIVES

After completing this experiment, you should be able to do the following:

1. Define qualitative analysis.
2. Test experimentally by quantitative analysis for lead, silver, and mercury ions.

APPARATUS

One-molar solution hydrochloric acid, 1.0-molar solution ammonium hydroxide, 1.0-molar solution nitric acid, potassium chromate, test solution containing Ag^+, Pb^{++}, and Hg^+ ions, ring stand, funnel, filter paper, one 250-mL beaker, three test tubes, hot water.

PROCEDURE

Use the flow diagram (Fig. 36.1) and proceed as follows:

1. Pour about 5 mL of the test solution (which contains lead, silver, and mercurous ions) into a clean, dry test tube.
2. Add four or five drops of hydrochloric acid.
3. Filter out the precipitate and discard the filtrate (the solution that goes through the filter paper).
4. Without removing the precipitates from the filter paper or funnel, wash thoroughly with hot water, using about half a test-tubeful, and collect the filtrate in a clean test tube.
5. Check for lead in the filtrate by adding potassium chromate. If lead is present, a yellow precipitate of lead chromate will be formed. *Note*: The yellow color will always be present. The solution will turn cloudy if lead is present.
6. After placing another test tube under the funnel, add ammonium hydroxide to the precipitate remaining on the filter paper. If mercury is present, the precipitate on the filter paper will turn black.
7. Check the filtrate for silver by acidifying it with nitric acid. The appearance of a precipitate (look carefully for it; the solution will show only a slight cloudiness) confirms the presence of silver.
8. Obtain from the instructor an unknown sample and, using the procedure given above, determine which of the above elements are present.

Number of unknown..   _____

Indicate the elements present or not present......................................   Pb_____

Hg_____

Ag_____

261

Figure 36.1 *Flow diagram. Separation of Ag⁺, Pb⁺⁺, and Hg⁺ in a known or unknown solution.*

Chemical Quantitative Analysis (Volumetric)

INTRODUCTION

Determination of the exact amounts of elements or substances present in an unknown sample is called quantitative analysis. The two most common methods of doing this are the **gravimetric method** and the **volumetric method**. The latter will be used in this experiment.

In the volumetric method, the solutions used in the reactions are of known strength. The volumes of these solutions needed to complete reactions are measured, and the quantities involved are calculated from the known strengths of the solutions. The completion of a reaction is made visible by the use of an indicator, a material added to the solutions. It shows by its color when the reaction is complete. The solutions are usually measured out of long, graduated tubes that have stopcocks at the end so that small quantities, even to a fraction of a drop, may be obtained as desired. Such tubes, called burettes, are commonly graduated to 0.10 mm.

All solutions used for testing in these experiments will be made up on the molar basis. *Thus, a one-molar solution will contain one gram-molecular weight of a substance per liter of solution. A liter is 1000 mL.* A 0.50-molar solution will contain 0.50 g molecular weight of the substance per liter of solution, and so forth.

LEARNING OBJECTIVES

After completing this experiment, you should be able to do the following:

1. Define quantitative analysis, molarity, and one-molar solution.
2. Calculate the percentage of acetic acid in vinegar.
3. Calculate the molarity of a solution.
4. Determine experimentally by quantitative analysis (volumetric method) the exact amount of a substance in an unknown sample.

APPARATUS

Two burettes, two Erlenmeyer flasks, phenolphthalein, 1.0-molar hydrochloric acid, 1.0-molar sodium hydroxide, vinegar, two 250-mL beakers.

PROCEDURE

1. Fifteen milliliters of vinegar (which includes acetic acid) will be placed in your Erlenmeyer flask together with two drops of phenolphthalein (an indicator that is colorless in acid solutions but turns red in basic solutions). About 50 mL of one-molar sodium hydroxide will be put in your burette. Read the level before you start. Now, from this burette slowly titrate sodium hydroxide to the flask containing the vinegar, keeping the solution well stirred, until the drop most recently added has caused the solution to turn a permanent pink. The appearance of this color indicates that the acid has just been neutralized by the base and the last drop has made the whole solution slightly basic. Read the burette carefully. Record the volume in Table 37.1.

   Because a one-molar solution contains one gram-molecular weight per liter, each milliliter contains one-thousandth of a gram-molecular weight. If, for example, a quantity of 17 mL of one-molar sodium hydroxide is required in the experiment above, then $17 \times 0.001 = 0.17$ (or 17/1000) of a gram-molecular weight of sodium hydroxide is required. From the equation

$$NaOH + HC_2H_3O_2 \rightarrow H_2O + NaC_2H_3O_2$$

   we know that one molecule of sodium hydroxide neutralizes one molecule of acetic acid, or in other words, one gram-molecular weight of sodium hydroxide neutralizes one gram-molecular weight of acetic acid.

   If, as in the above example, 17/1000 of a gram-molecular weight of sodium hydroxide is required, then 17/1000 of a gram-molecular weight is in the vinegar. The molecular weight of acetic acid is 60; therefore, the weight of acetic acid present in the vinegar sample is 17/1000 of 60 g, or 1.02 g.

   Calculate the percentage of acetic acid present in the vinegar sample that is provided for the experiment. Federal law requires a minimum of 4%.

2. The instructor will provide you with a known volume of hydrochloric acid. Slowly titrate sodium hydroxide to the volume of hydrochloric acid, and record the amount added in Table 37.2.

   Calculate the percentage of sodium hydroxide present in the solution supplied for you in the laboratory. The sodium hydroxide will be neutralized by a solution of hydrochloric acid whose strength will be given to you in the laboratory. The equation is

$$NaOH + HCl \rightarrow H_2O + NaCl$$

Data Table 37.1

Volume of vinegar	15 mL
Volume of sodium hydroxide	_____ mL
COMPUTATIONS Weight of volume of vinegar	15 g
Weight of acetic acid present	_____ g
Percentage of acetic acid	_____ %

Data Table 37.2

Volume of hydrochloric acid	15 mL
Volume of sodium hydroxide solution	_____ mL
COMPUTATIONS Weight of volume of NaOH solution	_____ g
Weight of NaOH present	_____ g
Percentage of NaOH	_____ %

QUESTIONS

1. Find the number of grams of KCl in 500 mL of a 3-molar solution.

2. What would be the molarity of an NaCl solution if 250 g of salt is used to make 5 L of solution?

Experiment 38

Kepler's Law

INTRODUCTION

In this experiment, we investigate the effects of a central force on the motion of an object. A central force may be defined as one that is always directed toward some particular point. The gravitational attraction between Earth and the Sun is an example of an inverse-square central force (Fig. 38.1). The path resulting is an ellipse. All central forces may produce elliptical paths. In the seventeenth century, Kepler discovered several properties of these ellipses. We investigate one of these, the law of equal areas, which may be formally stated as follows: When an object is traveling in an ellipse under the action of a central force, a line drawn from the force center to the object sweeps out equal areas in equal times.

The drawing in Figure 38.2 illustrates the path of an object traveling in an ellipse under the action of a central force. The time interval between the dots shown on the diagram is 0.1 second.

LEARNING OBJECTIVES

After completing this experiment, you should be able to do the following:

1. State Kepler's law of equal areas.
2. Determine experimentally the validity of the law of equal areas.

APPARATUS

Modified drawing of a photograph of the path of a pendulum bob, rulers, protractors.

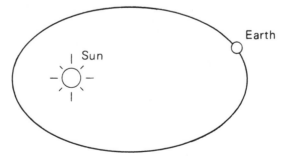

Figure 38.1 *Earth revolves around the Sun in an elliptical path, with the Sun at one focus.*

PROCEDURE

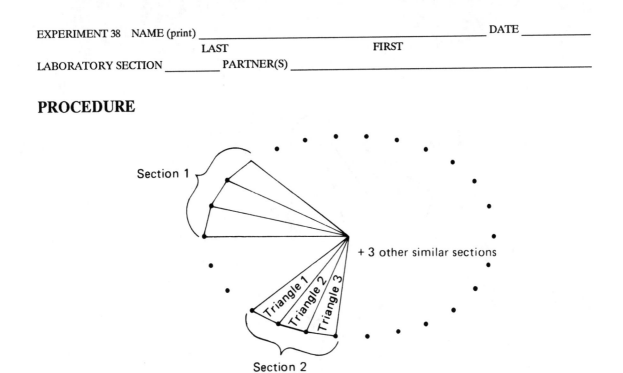

Figure 38.2

Divide your copy of the drawing of the photograph into sections containing three time intervals each. (See Fig. 38.2.) According to Kepler's law, the area of each of these sections should be equal. By the formula below, obtain the area of several of these sections and record them in Data Table 38.1. From your table, find the average area, and the percent difference. (Use the largest deviation from the average in this last calculation.)

To find the area of a triangle, we use the relation:

$$\text{Area of triangle} = 1/2 \text{ width} \times \text{height}$$

or

$$A = 1/2Wh$$

where A = area,

  W = width,

  h = height.

To use this formula to find the area of the triangles formed on the drawing, we shall follow the procedure outlined here. See Fig. 38.3.

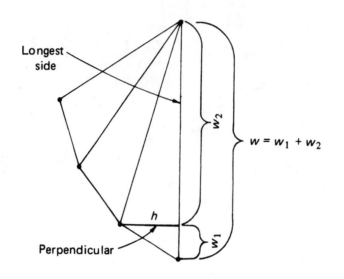

Figure 38.3

1. Draw lines from the force center to each of four points in a row. (Choose the four consecutive points closest to the force center.)
2. Draw lines connecting the four points chosen. (This will give three triangles side by side.)
3. Construct a perpendicular from the *longest* side of each triangle to the dot opposite it.
4. Carefully measure the length of the perpendicular h and the long side W. Be as accurate as possible in this measurement.
5. Find the area of the first triangle using the formula for the area given above.
6. Do the same for the other two triangles in this section. Call it Section 1.
7. Now add these three areas together to get the total area of Section 1. Record data in Data Table 38.1.
8. Now do four other sections (each consisting of three triangles) and record all data in Data Table 38.1.
9. Find the average area of the sections measured and find the largest percent difference between your calculated areas and the average.

Data Table 38.1

Section number	Area of triangle$_1$	Area of triangle$_2$	Area of triangle$_3$	Total area of section
1				
2				
3				
4				
5				

Average area...   _____

Percent difference..   _____
(Show your work.)

QUESTIONS

1. From your observation of the drawing of the photograph, state where you believe the speed to be greatest and why.

2. Determine the period (time of one revolution) of the orbiting body.

3. Where in Earth's orbit around the Sun is the speed of Earth the greatest?

4. Refer to your textbook, if necessary, and name the month in which Earth's orbital speed is the greatest.

Figure 38.4 *Drawing showing the elliptical path of a revolving body around a central force.*

Stars and Their Apparent Motions

INTRODUCTION

The best way to study the stars and their motions would be to observe them night after night for a year or more. Because weather conditions would not permit this, and the astronomy laboratory for this course is for only two periods, other methods must be used. The next best method would be to use a planetarium, where the observer is placed at the center of a large half-sphere or dome. The stars and their apparent motions are projected by optical means on the dome. However, the method we shall use in this experiment will be that of the celestial sphere. A celestial sphere is a small model sphere (they are made in several sizes) that may be set to portray the sky as seen from any latitude on Earth, on any day and for any time of day.

LEARNING OBJECTIVES

After completing this experiment, you should be able to do the following:

1. Define the terms celestial latitude, celestial longitude, altitude, declination, ecliptic, vernal equinox, zenith, perpetual apparition, and perpetual occultation.

With the aid of the celestial sphere:

2. Determine the declination of the Sun on any day of the year.
3. Determine the time of sunrise and sunset for any latitude on any day of the year.
4. Determine the approximate number of hours of daylight and darkness from any latitude.
5. Determine the altitude of the Sun at 12 noon local solar time for any latitude on any day of the year.
6. Identify and give the location and apparent motion of the brighter stars and constellations.

DISCUSSION

The apparent position of a star changes for an observer as the observer changes latitude. For example, Polaris, the north star, appears directly overhead for an observer at 90° N. When the observer is at 0° latitude, the north star will appear on the northern horizon. If the observer travels northward 1°, the north star will appear 1° above the northern horizon. At 40° N, the north star will appear 40° above the northern horizon. If the observer travels south of the equator 1°, then the north star will be 1° below the northern horizon and will not be visible to the observer. Thus, a change in latitude by the observer makes an apparent change in the positions of the stars in respect to the observer.

To understand how the celestial sphere is used to view the positions of the stars, place the celestial sphere in front of you as you read this discussion, and locate each item on the model as it is defined. Refer also to Fig. 39.1 and Fig. 39.2.

The celestial sphere represents the sky, not Earth. Earth is a point located at the center of the sphere. As viewed from above the north pole, Earth turns counterclockwise or eastward. The eastward rotation of Earth about its axis produces the apparent westward motion of the celestial sphere. Thus, when you rotate the celestial sphere westward the rotation represents an apparent, not actual, motion. For example, we observe the Sun to rise on the eastern horizon, travel westward across the sky, and set in the west. This apparent motion of the Sun is due to Earth's rotation about its axis.

Figure 39.1 *Celestial sphere, a model sphere portraying the sky. The celestial equator is the line circling the sphere midway between the north and south celestial poles. The ecliptic plane is at an angle 23½° to the celestial equator. Earth is located at a point in the center of the sphere. (Courtesy of Dennoyer-Geppert Company)*

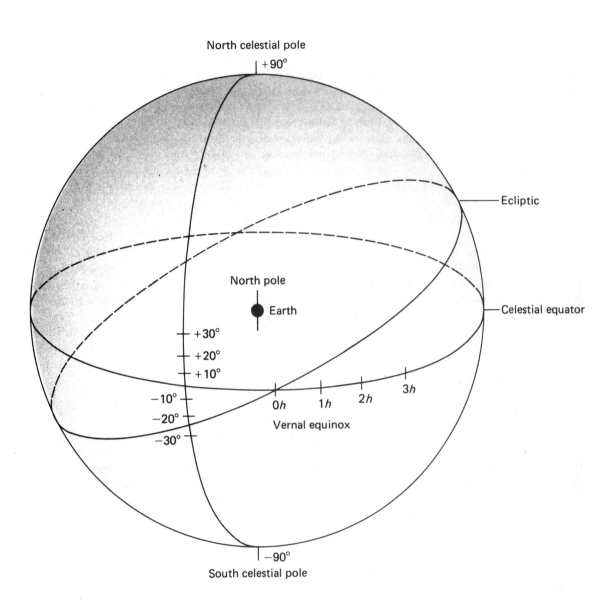

Figure 39.2 *Diagram illustrating declination and right ascension. Declination is from 0° to +90° for the northern part of the celestial sphere, and from 0° to −90° for the southern part. Earth is represented by the small dot at the center of the celestial sphere.*

The large wooden ring constructed around the sphere, parallel to the floor, represents the observer's horizon. If you are outdoors, the horizon appears as the dividing line between Earth and the sky. The horizon appears 90° down from directly overhead (observer's zenith), and extends completely around the observer a full 360°. **Zenith** is the point on the celestial sphere opposite the direction of gravity. Only those stars above the horizon (the wooden ring) can be seen by the observer. The metal ring positioned 90° to the horizon, and circling vertically over the sphere, represents the observer's overhead meridian.

The position of a star on the celestial sphere is given in respect to two reference frames. The first is the **celestial equator,** an imaginary line circling the celestial sphere halfway between the north and south celestial poles. If Earth's equator were extended outward, it would pass through the celestial equator. That is, Earth's equator and the celestial equator are in the same plane. **Celestial latitude** is stated in respect to the celestial equator and is similar to latitude measurements for the surface of the Earth, except that degrees toward the north celestial pole are labeled plus (+) and degrees toward the south celestial pole are labeled negative (−). The second reference frame is the **celestial prime meridian.** This is an imaginary line running from the north celestial pole to the south celestial pole, perpendicular to the equator, and intersecting the celestial equator at the point of the vernal (spring) equinox on the ecliptic. See Fig. 39.2. The **vernal equinox** is the point on the ecliptic where the Sun crosses the celestial equator from south to north. The **ecliptic** is the apparent annual path of the Sun on the celestial sphere. **Celestial longitude** is stated in respect to the vernal equinox and is measured to the east along the celestial equator from the vernal equinox. The vernal equinox is taken as the starting point and labeled zero (0) hour.

The angular distance north or south of the celestial equator is called **declination.** The range of declination is from 0° to +90° for the northern part of the celestial sphere, and from 0° to −90° for the southern part of the celestial sphere. The angular distance measured eastward from the vernal equinox is called **right ascension,** and is measured in hours, minutes, and seconds. The range of right ascension is from 0 h to 24 h. For example: The star Arcturus has a declination of −20° and a right ascension of 14 h, 16 m. Similar coordinates are given for the location of any star.

The celestial sphere is used to portray the celestial sky for an observer located at any latitude on the earth, on any day of the year, at any time of day. To set the celestial sphere for a latitude, a day, and a time of day, proceed as follows:

Step 1 To set the sphere for a particular latitude, use the position of the north star in respect to the northern horizon. If you are in the northern hemisphere of Earth, the north star will be visible. Therefore, set the north star at an altitude equal to your latitude. **Altitude** is the angle between the line of sight to the star (in this case, the north star) and the horizon. To do this, move the sphere by moving the brass ring, which supports the sphere, up or down until the latitude of the star is equal to your latitude. *Be sure to count the number of degrees on the brass ring between the north star and the northern horizon.* The reading on the meridian may be correct, or it may be the complementary angle. To be sure, count the degrees. To set the sphere for a southern hemisphere latitude, place the north star below the northern horizon, with the number of degrees equal to the observer's southern latitude. Again, count the number of degrees to be sure the setting is correct.

Step 2 By definition, when the Sun crosses the local overhead meridian in its daily apparent motion across the sky, it is 12 noon, local solar time. Hence, to portray the sky as seen at 12 noon, it is necessary to locate the Sun on the ecliptic, then turn the sphere until the Sun is at the overhead meridian. To do this, find the day of the year as marked on the celestial equator. Place this date under the overhead meridian. This automatically places the Sun under the overhead meridian. Remember, the sun is on the ecliptic, not the celestial equator. The Sun will cross the celestial equator twice each year, once apparently going northward (vernal equinox) and again apparently going southward (autumnal equinox).

Step 3 Once the position of the sphere is set for 12 noon on a particular day, it is easy to set the sphere to correspond to any other hour of the same day by rotating the sphere westward 15° for each hour past noon, or eastward 15° for each hour before noon. Thus, any hour of the day can be established on the sphere. Because there are 60 min in 1 h and 15° represents 1 h on the celestial sphere, each degree represents 4 min of time. (Sixty min divided by 15° = 4 min/degrees.) Thus, the celestial sphere can be set approximately for minutes.

Once the celestial sphere has been set for a particular latitude, day, and time of day, everything above the horizon (the wooden ring) is visible to the observer. If it is nighttime, the stars can be seen. If it is daytime, the stars are there but the brightness of the sunlight prevents them from being seen.

APPARATUS

Celestial sphere and one small, ¼-in, circular disk, cut from masking tape (yellow preferred) to be used to represent the Sun's position on the ecliptic.

PROCEDURE 1

Place the celestial sphere on the laboratory table so that the overhead meridian is in a north-south direction, with the north pole of the sphere in the general direction of geographic north. The word "north" appears on the north wall of the laboratory. Record the date of this experiment. _____

1. Set the celestial sphere to portray the sky as seen by an observer located at Earth's equator (0° latitude). Place Polaris on the northern horizon. Rotate the globe until the date you recorded above appears under the overhead meridian. This sets the celestial sphere for 0° latitude and 12 noon local solar time on this date. Stick the ¼-in circular disk of masking tape on the ecliptic where it passes under the overhead meridian. This represents the Sun and its position on the ecliptic.

   Determine the declination of the Sun.

   Declination of the Sun..   _____

   Determine the right ascension of the Sun.

   Right ascension of the Sun...   _____

   Determine the altitude of the Sun at 12 noon local time.

   Altitude of the Sun...   _____

   Determine the altitude of the north star (Polaris) at this time.

   Altitude of Polaris..   _____

2. Set the celestial sphere for 4 P.M. local time for the latitude and date given above. This is done by rotating the sphere westward 4 h or 60° from the 12 noon position. *Note*: Westward is to your left, if you are facing north. Look for the sign (north) on the wall of the laboratory. Also, the word (west) or W may be printed on the horizon ring.

   Determine the approximate altitude of the Sun at this time. State your answer in degrees above or below the western or eastern horizon. Example: 16° above western horizon.

   Altitude of the Sun...   _____

   Altitude of Polaris at this time ....   _____

3. Set the celestial sphere to portray the sky as seen by an observer located at the north pole (90° N). Rotate the globe until the date recorded above appears under the overhead meridian. This sets the celestial sphere for 90° N and 12 noon local solar time on this date. Stick the ¼-in circular disk of masking tape on the ecliptic where the ecliptic passes under the overhead meridian. This represents the Sun and its position on the ecliptic.

   Determine the altitude of the Sun at 12 noon local time.

   Altitude of the Sun...   _____

   Altitude of Polaris at this time ....   _____

4. Rotate the sphere westward until 8 P.M. is represented by the celestial sphere.

   Altitude of the Sun...   _____

   Altitude of Polaris at this time ....   _____

5. Set the sphere to portray the sky as seen from Washington, D.C. (39° N), on the date of this experiment or use the latitude of your local city or town. Determine the altitude of the Sun at 12 noon local time. Use the circular disk to represent the Sun.

Altitude of the Sun .... _____

Altitude of Polaris at this time .... _____

Determine the local time of sunrise and sunset. Sunrise occurs when the Sun is on the eastern horizon. Sunset occurs when the Sun is on the western horizon. Proceed as follows to determine the time of sunset. Place your forefinger on the Sun, which should be under the overhead meridian. The time is 12 noon local mean solar time. Rotate the sphere westward counting the hours (the hours are marked on the celestial equator) as they pass under the overhead meridian. Rotate the sphere until your finger touches the western horizon. The time of sunset is the number of hours you counted plus any minutes past 12 noon. To determine the time of sunrise proceed as follows: Place your forefinger on the Sun, which should be under the overhead meridian. Rotate the sphere eastward counting the hours as they pass under the overhead meridian. Rotate the sphere until your finger touches the eastern horizon. The time of sunrise is the number of hours counted plus any minutes subtracted from 12 noon.

Time of sunrise .... _____

Time of sunset .... _____

How many hours of daylight did Washington, D.C., or your home city have on the date of this experiment? .... _____

How many hours of darkness? .... _____

How many hours of daylight did the equator (0° latitude) have on this date? (Determine the time of sunrise and sunset.) .... _____

How many hours of daylight did the north pole have on this date? .... _____

How many hours of daylight did the south pole have on this date? .... _____

State the regions of perpetual apparition and perpetual occultation. That is, state the regions on the celestial sphere that always appear (regions that never go below the horizon to an observer at 39° N latitude), and the regions that never appear (regions that never come above the horizon to an observer at this latitude). Rotate the sphere 360° and observe the regions that stay above and below the horizon. The answers will be in degrees. Example: 52° north to 90° north.

Apparition .... _____

Occultation .... _____

6. On June 21 and December 21 the Sun will be at its maximum declination north and maximum declination south, respectively. Set the celestial sphere to portray the sky from 39° N. Determine the declination of the Sun on June 21 and December 21. Determine the altitude of the Sun on these dates and the degrees east of due south where the Sun rises on these two dates, and the degrees west of due north where the Sun sets on these two dates.

Declination of the Sun (June 21) .... _____

Declination of the Sun (December 21) .... _____

Altitude of the Sun on June 21 .... _____

Degrees east of due south for sunrise on June 21 .... _____

Degrees west of due south for sunset on June 21.. _____

Altitude of Sun on December 21 .... _____

Degrees east of due south for sunrise on December 21 .... _____

Degrees west of due south for sunset on December 21...................................... _____

PROCEDURE 2

1. Set the celestial sphere to portray the sky as seen from Washington, D.C., or your local city on the date of this experiment at 9:00 P.M., local solar time. Show the location of the Little Dipper, Big Dipper, and Cassiopeia by drawing them on the diagram in this experiment. Remember, the observer is located at the center of the celestial sphere looking up and northward to see these constellations. Check your drawing by observing these constellations at 9:00 P.M., local solar time, on the date of this experiment.

2. Determine the local solar time at which the Great Nebula in Andromeda (designated M31 on most celestial spheres) is on the overhead meridian. If your model does not show this notation, pick one of the brightest stars in this constellation. .... _____

3. State the declination and right ascension of the Galaxy M31 in Andromeda.

   Declination.. _____

   Right ascension.. _____

   Note: M31 can be seen with the unaided eye on a clear night. Check the celestial sphere for the position of M31, then observe on a clear night.

4. The brightest star in the sky is Sirius. At what local solar time will Sirius rise on this day at Washington, D.C., or your local city?... _____

5. At what local solar time will Sirius be on the overhead meridian? .... _____

6. What is the altitude of Sirius at the local solar time of crossing the overhead meridian? _____

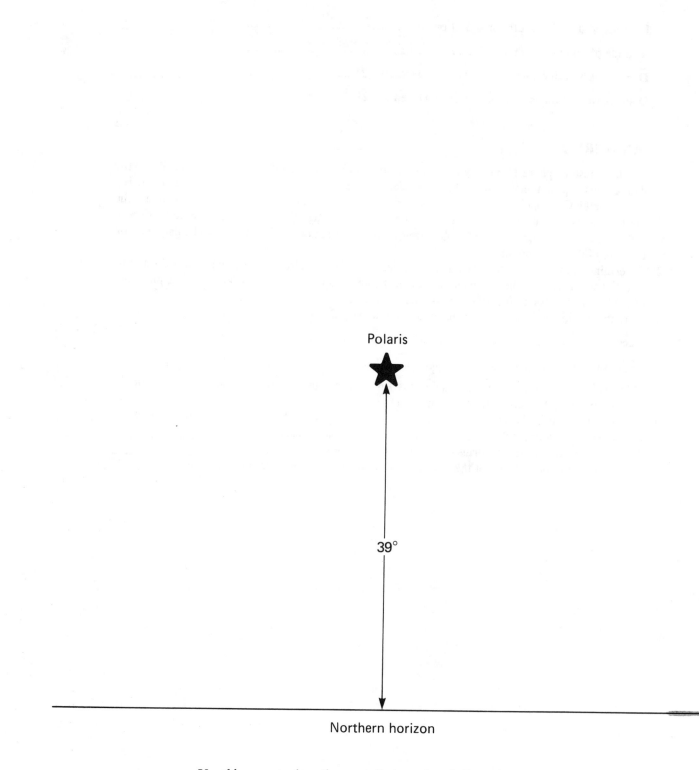

Polaris

39°

Northern horizon

Use this space to draw the constellations given in Procedure 2, Step 1.

QUESTIONS

1. How does the declination of the Sun vary over one year?

2. Where on the ecliptic is the vernal equinox?

3. How does the altitude of the north star vary for an observer located at the equator? At 39°N?

4. Why was the globe rotated westward to position the celestial sphere to portray a P.M. local time?

5. Does the Sun ever set north of due west for an observer located at Washington, D.C. (39°N)?

Locating Stars in the Night Sky

INTRODUCTION

This experiment is a combination of in-lab work that will teach you about the stars in the night sky that can be observed during this season of the year and an actual observation of the sky that can be made at a "star party" held by your instructor or as an individual homework assignment. Experiment 39 is also about the stars. It explains the use of a celestial sphere. You can refer to Experiment 39 if you are not familiar with how to use a celestial sphere when you do step 2 of this experiment. If a celestial sphere is not available, you can get the information needed for Data Tables 40.2 and 40.3 from a star atlas or by using handouts available from your instructor.

LEARNING OBJECTIVES

After completing this experiment, you should be able to do the following:

1. List at least six constellations that are visible at this season of the year.
2. Find constellations and stars on a celestial sphere and determine their right ascension, declination, and zenith angles.
3. Explain how to use a star atlas to find information about the stars.
4. Locate the prominent constellations and stars in the sky at this time of year and identify them.

APPARATUS

Star Charts 40.1–40.4, blank Star Charts 40.5–40.8, star atlas or handouts from instructor, celestial sphere.

PROCEDURE

1. Select the star chart provided that represents the current season of the year. Study it for about 10 minutes to get to know the location of the constellations that are now present in the night sky.

2. Take one of the blank star charts and fill in as much information about the night sky as you can remember from your previous 10-minute study. Do not refer back to the "full" star chart. You will get to correct any mistakes or omissions in the next step. Be as neat and as accurate as you can.

3. Now using the "full" star chart, fill in all of the details that you missed in step 2. This procedure is designed to prepare you to successfully locate the principle stars in the night sky when you get outside under the stars.

4. Fill in the information in Data Table 40.2 for the stars listed in Data Table 40.1 that appear in the current night sky. Use the celestial sphere to find this data. Set the celestial sphere for 9:00 P.M. local solar time on today's date and a location corresponding to your latitude. Use 40°N if you do not know your latitude. (If you do not have a celestial sphere, skip this step and go on to step 5. You can fill in the data in Data Table 40.2 from a star atlas or from handouts passed out by your instructor.)

Data Table 40.1

This data table lists the stars to use in steps 4 and 5 for the various seasons of the year. You will be able to see those indicated for the current season when you actually perform sky observations later in this experiment (step 6).

	Fall	Winter	Spring	Summer
1	Aldeberan	Aldeberan	Arcturus	Altair
2	Algol	Betelgeuse	Capella	Antares
3	Altair	Capella	Castor	Arcturus
4	Capella	Castor	Polaris	Deneb
5	Deneb	Polaris	Pollux	Polaris
6	Formalhaut	Pollux	Procyon	Regulus
7	Polaris	Regulus	Regulus	Spica
8	Vega	Rigel	Sirius	Vega

Data Table 40.2

Star name	Constellation	Right ascension	Declination	Altitude	Zenith angle	Compress direction
1						
2						
3						
4						
5						
6						
7						
8						

Note: Altitude + zenith angle = 90°

5. Using one of the star atlases, look up the stars used in Table 40.2. Check the right ascension and declination values to make sure the ones you found on the celestial sphere are correct and fill in the other data in Data Table 40.3.

Data Table 40.3

Star name	Right ascension	Declination	Apparent magnitude	Luminosity compared to Sun	Spectral class
1					
2					
3					
4					
5					
6					
7					
8					

6. Fill in the names of the stars and the constellations in Data Table 40.4 using Data Table 40.1 and the appropriate seasonal star chart. Then go outside and look at the stars in the night sky some evening when the weather is clear. Fill in the remainder of the data in Data Table 40.4 from your own personal observations. You will not need a telescope; you can see all of these stars with the unaided eye on a clear night.

Data Table 40.4

Star name	Constellation	Approximate zenith angle	Approximate compass direction	Apparent color of star	Apparent magnitude of star
1					
2					
3					
4					
5					
6					
7					
8					

7. Your instructor will tell you if any planets are in position to be observed at this time. If there are any planets visible, you can observe them as you study the stars. Answer the following questions.

   1. What planet(s) are visible tonight? _____

   2. In which constellation(s)? _____

   3. Your estimate of planet(s) brightness: _____

   4. Apparent color of planet(s): _____

Star Chart 40.1   (Full)
Sky as Seen from 40°N Latitude

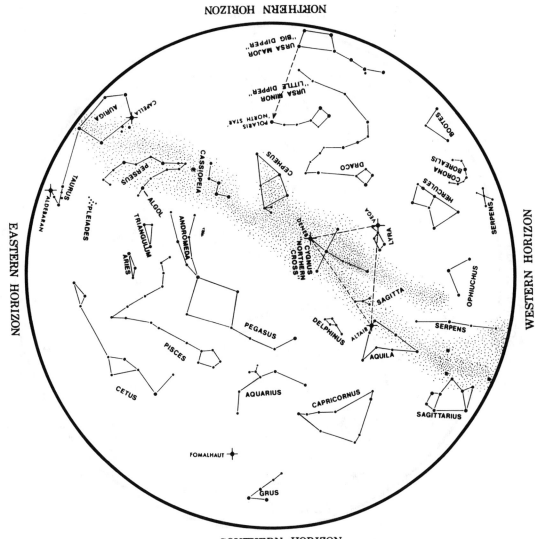

The Night Sky in Fall
(September, October, November)
11 P.M.   9 P.M.   7 P.M.

Star Chart 40.2 (Full)
Sky as Seen from 40°N Latitude

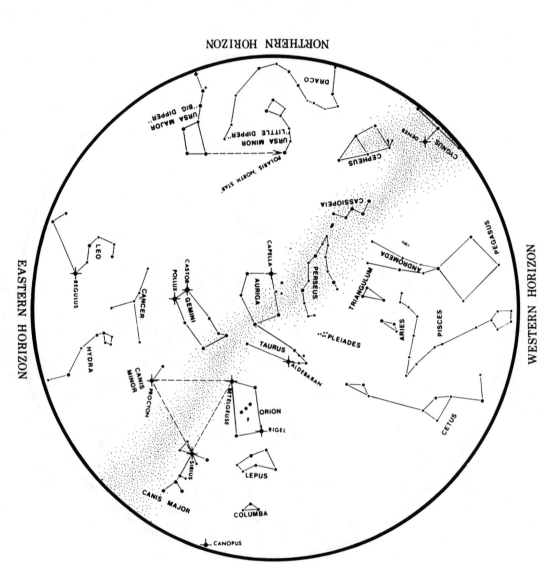

NORTHERN HORIZON

EASTERN HORIZON

WESTERN HORIZON

SOUTHERN HORIZON

The Night Sky in Winter
(December, January, February)
11 P.M.   9 P.M.   7 P.M.

Star Chart 40.3   (Full)
Sky as Seen from 40°N Latitude

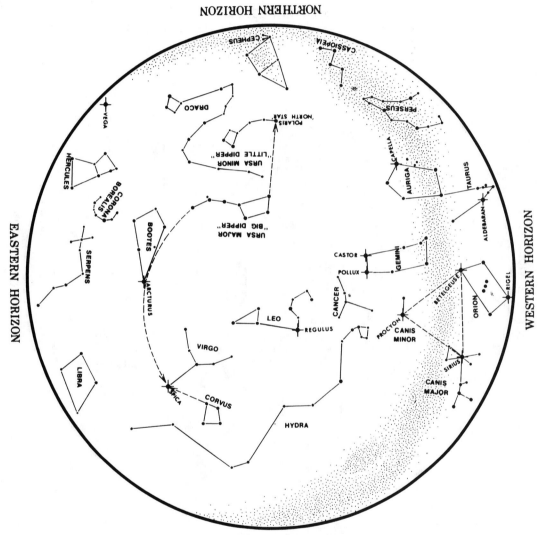

The Night Sky in Spring
(March, April, May)
11 P.M. 9 P.M. 7 P.M.

The Night Sky in Summer
(June, July, August)
11 P.M. 9 P.M. 7 P.M.

Star Chart 40.5   (Blank)
Sky as Seen from 40°N Latitude

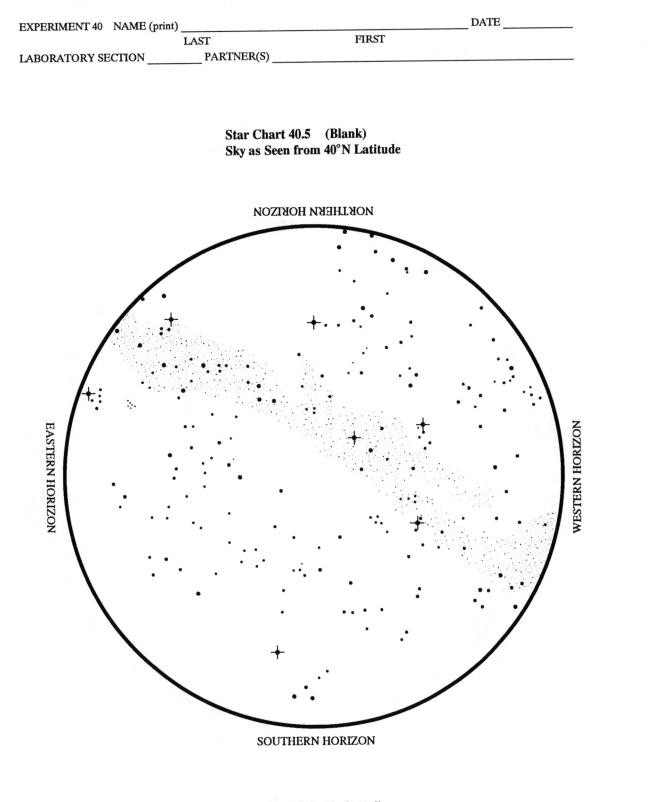

NORTHERN HORIZON

EASTERN HORIZON

WESTERN HORIZON

SOUTHERN HORIZON

The Night Sky in Fall
(September, October, November)
11 P.M.   9 P.M.   7 P.M.

Star Chart 40.6   (Blank)
Sky as Seen from 40°N Latitude

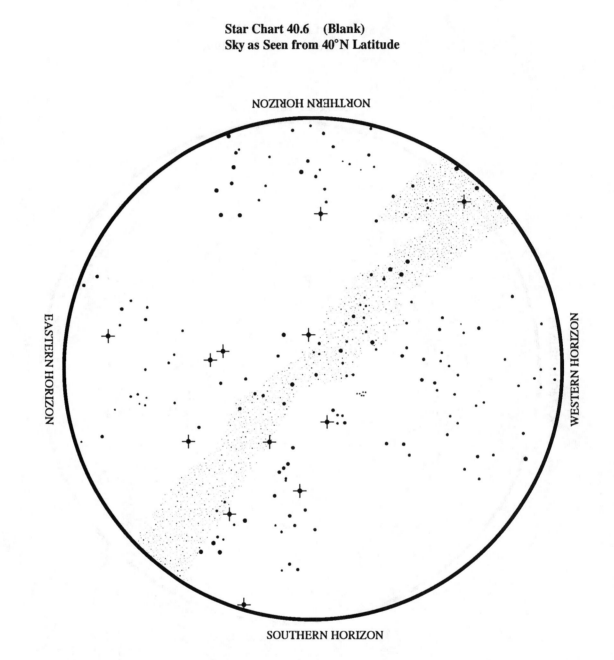

NORTHERN HORIZON

EASTERN HORIZON

WESTERN HORIZON

SOUTHERN HORIZON

The Night Sky in Winter
(December, January, February)
11 P.M.   9 P.M.   7 P.M.

Star Chart 40.7   (Blank)
Sky as Seen from 40°N Latitude

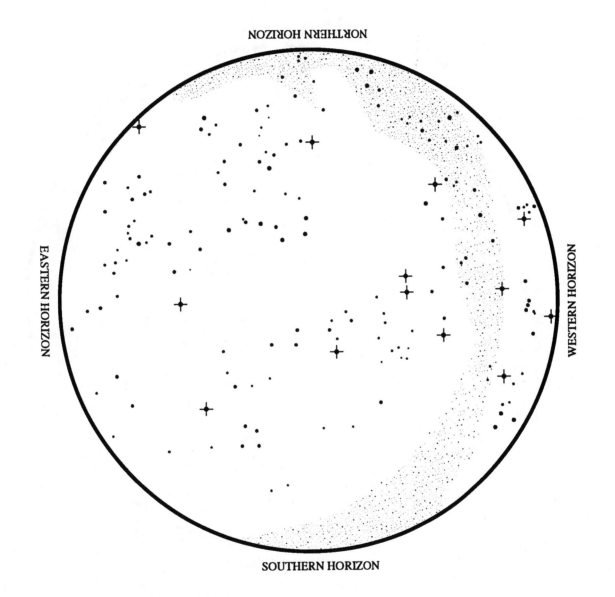

The Night Sky in Spring
(March, April, May)
11 P.M. 9 P.M. 7 P.M.

      297

Star Chart 40.8   (Blank)
Sky as Seen from 40°N Latitude

NORTHERN HORIZON

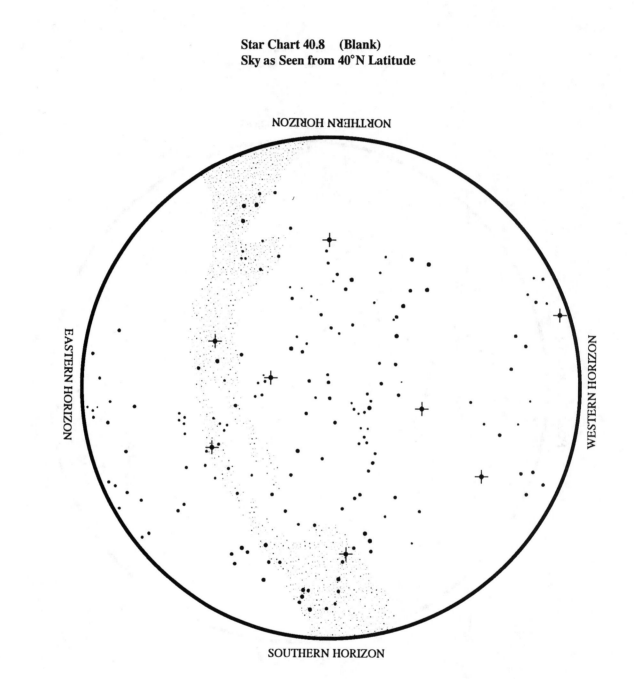

EASTERN HORIZON

WESTERN HORIZON

SOUTHERN HORIZON

The Night Sky in Summer
(June,   July,   August)
11 P.M. 9 P.M. 7 P.M.

Motions and Phases of the Moon

INTRODUCTION

The moon is a satellite of Earth that makes one revolution eastward around Earth every 29½ solar days. The orbit is that of an ellipse inclined 5° to the ecliptic and crossing the ecliptic twice each revolution.

By definition, when the moon is on the meridian with the Sun, it is in new phase. Since the dark side of the moon is toward Earth at this time, the new moon cannot be seen by the observer on Earth (see Fig. 41.1). When the moon is 90° east of the Sun, the moon is in the first-quarter phase. At this time it will appear on the meridian at 6:00 P.M., local solar time, in the form of one-half of a full circle. As the moon revolves eastward, it appears larger in face size until at 180° east of the Sun, the moon is in full phase, and appears on the meridian at 12 midnight, local solar time, as a full disk. Another 90° eastward will locate the moon 270° east of the Sun, or 90° west of the Sun. The moon now appears on the meridian at 6:00 A.M., local solar time, as a quarter moon. This phase is known as third-quarter or last-quarter moon.

Although the moon's orbit is inclined at an angle of 5° with the ecliptic plane, for simplicity in this experiment we shall assume that the moon travels in the ecliptic plane.

LEARNING OBJECTIVES

After completing this experiment, you should be able to do the following:

1. State the different phases of the moon and the local times when they occur.
2. With the use of the celestial sphere determine the phase of the moon, and its approximate altitude, when the time of the meridian crossing is known.

APPARATUS

Celestial sphere and a small circular disk cut from masking tape as called for in Experiment 39 plus a half-circular disk of the same diameter for representing the moon's position on the ecliptic.

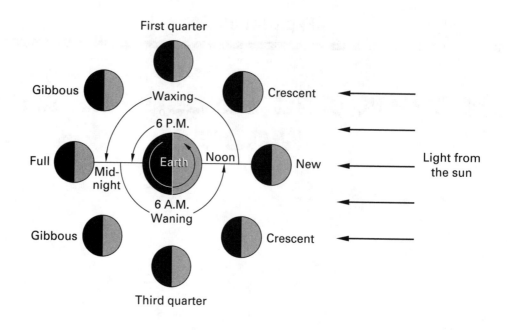

Figure 41.1 *Phases of the moon as observed from a position in space above the north pole of Earth.*

PROCEDURE

1. Set the celestial sphere to portray the sky as seen from Washington, D.C. (39° N), or from your local city on December 22. Use the two pieces of masking tape to make the positions of the Sun and moon in the following procedure:

   (a) Using the celestial sphere, determine the altitude of the new moon as observed from Washington or your local city, if the moon was in the new phase on December 22.

   Altitude (moon on overhead meridian) .... _____

   (b) Seven and three-eighths days later, the moon would be in the first-quarter phase. For simplicity, assume that the Sun remains at the same declination (23½° S) and determine the altitude, rising time, and setting time of the first-quarter moon.

   Altitude (moon on overhead meridian) .... _____

   Local solar time moon rises.. _____

   Local solar time moon sets .... _____

   (c) At the end of another 7 3/8 days, the moon will be in the full phase. Again, in order to make solving the problem easier, assume that the Sun remains at the same declination. Determine the altitude, rising time, and setting time of the full moon.

   Altitude (moon on overhead meridian) .... _____

   Local solar time moon rises.. _____

   Local solar time moon sets .... _____

   (d) As the moon revolves eastward, it will be in the third-quarter or last-quarter phase another 7 3/8 days later. Assuming that the Sun remains at 23½° S, find the altitude, rising time, and setting time of the last-quarter moon.

   Altitude (moon on overhead meridian) .... _____

   Local solar time moon rises.. _____

   Local solar time moon sets .... _____

2. Obtain from the instructor in charge of the laboratory the date in the current month when the new moon occurs. (Use this date throughout this question.)

   Date of the new moon.. _____

   (a) Using the celestial sphere, determine the altitude of the new moon as observed from Washington or your local city.

   Altitude (moon on overhead meridian) .... _____

   Note: For this series of problems, we shall not assume that the Sun remains at the same declination, as we did in Procedure 1. The Sun appears to move about one degree per day. Allow for this movement as you solve the following problems. To find the declination of a particular phase of the moon, remember its position in respect to the Sun, as given in the introduction to this experiment.

(b) Seven and three-eighths days after the date of the new moon, the moon will be in the first-quarter phase. Solve for the following information, using previously learned facts and the celestial sphere.

Date of first-quarter moon .... _____

Local solar time first-quarter moon is on overhead meridian .... _____

Standard time first-quarter moon is on overhead meridian................................ _____

Altitude of first-quarter moon when on overhead meridian............................... _____

Rising time (local solar) of first-quarter moon... _____

Rising time (standard) of first-quarter moon... _____

Setting time (local solar) of first-quarter moon.. _____

Setting time (standard) of first-quarter moon.. _____

(c) With the passing of another 7 3/8 days, the moon will be in full phase. Solve for the following information using previously learned facts and the celestial sphere.

Date of full moon.. _____

Local solar time full moon is on the overhead meridian .... _____

Standard time full moon is on the overhead meridian....................................... _____

Altitude of full moon when on overhead meridian.. _____

Rising time (local solar) of full moon .... _____

Rising time (standard) of full moon.. _____

Setting time (local solar) of full moon .... _____

Setting time (standard) of full moon... _____

Degrees east of due south at which full moon rises .... _____

Degrees west of due south at which full moon sets.. _____

(d) When the moon has revolved to 270° east of the Sun, it will be in the last-quarter phase. Solve for the following information using previously learned facts and the celestial sphere.

Date of last-quarter moon .... _____

Local solar time last-quarter moon is on the meridian...................................... _____

Standard time last-quarter moon is on the meridian .... _____

Altitude of last-quarter moon when on overhead meridian .... _____

Rising time (local solar) of last-quarter moon.. _____

Rising time (standard) of last-quarter moon... _____

QUESTIONS

1. During what month will the full moon achieve its maximum altitude, as observed from Washington, D.C. (39°N)? minimum altitude?

2. Is it possible to see a waxing crescent moon on the overhead meridian at 10 A.M. local time? at 4 P.M. local time? Explain your answers.

3. In the experiment, what is the maximum possible error in degrees that could be made in determining the moon's altitude by assuming the moon traveled in the same plane as the ecliptic?

4. Assuming the moon actually travels in the same plane as the ecliptic, how many total lunar eclipses would occur each year? How many total solar eclipses?

Observing the Phases of the Moon

INTRODUCTION

This experiment is a long-term project in which you will observe the phases of the moon and keep a log of its position in the sky along with the data and time of your observations. To best study the complete phase cycle of the moon, your observations should be made over at least four weeks, but the work can be done in as short a period as two weeks if time is limited.

Experiment 41 also deals with the phases of the moon, but uses a celestial sphere for gathering the data. You may be assigned to do Experiment 41 or at least look it over before you complete your work on this experiment. The two experiments complement each other nicely; this one is based on actual observation of the moon in the sky, while the previous one is an in-class study of the same general material.

The moon revolves around Earth once every 27.3 sidereal days, but because Earth is also moving in its orbit around the Sun, the cycle of the phases of the moon as seen from Earth is about two days longer than this. The lunar phases follow a cycle called the *synodic period*, which is 29.5 days. The moon's orbit is inclined five degrees to the plane of the ecliptic so that even though the moon's path crosses the ecliptic plane twice each revolution, eclipses of the moon and Sun do not occur every month.

The moon's surface is illuminated by sunlight and one-half of the lunar surface is illuminated at all times. The fractional portion of this lighted hemisphere that can be seen from Earth depends on the relative positions of the Sun, Earth, and moon when the observation is being made. As this fractional portion of the moon's lighted surface changes over time, an observer on Earth sees the monthly cycle of lunar phases.

When the moon is on your overhead meridian at the same time as the Sun, the moon is in the *new phase*. This occurs at 12 o'clock noon local solar time (LST). Because the dark side of the moon is toward Earth and the Sun is at the same location in the sky at that time, the moon is quite difficult to observe.

Over the next 7 3/8 days, the position of the Sun and the moon will shift relative to Earth in such a way that an increasing amount of the lighted portion of the lunar surface will become visible from Earth. At first only a small crescent of the lighted surface can be seen, but this steadily increases until half of the surface facing Earth is lighted. During this week, the moon is said to be in the *waxing crescent phase*.

When the moon is 90 degrees east of the Sun, one-half of its illuminated surface is visible, and the moon is then said to be in the *first-quarter phase*. At this time, it will appear on an observer's overhead meridian at 6:00 P.M. LST. After the first-quarter phase, the lighted portion of the lunar surface that is visible from Earth continues to increase over the next 7 3/8 days and the moon is said to be in the *waxing gibbous phase*.

When the entire hemisphere of the moon facing Earth is lighted, the observer will see a *full phase*. This happens when the moon reaches a position 180 degrees east of the Sun and appears on your overhead meridian at 12 o'clock midnight LST as a fully illuminated disk. After full phase, the portion of lighted lunar surface as seen from Earth decreases. This portion of the monthly lunar cycle is called the *waning gibbous phase*.

At 270 degrees east of the Sun (or 90 degrees west of the Sun), the moon appears on the meridian at 6:00 A.M. LST, and has reached the *third-quarter* or *last-quarter phase*. At this time, one-half of its lighted surface is visible again from Earth. As the moon continues its eastward motion, it will eventually be on the overhead meridian at the same time as the Sun and thus return to the *new moon phase*. During the period between the third-quarter and new phase, the lighted portion of the moon visible from Earth is still decreasing, and this portion of the cycle is called the *waning crescent phase*.

The moon is only in its new, first-quarter, full, and last-quarter phases for an instant. These are defined positions relative to the Sun and Earth. Therefore, the moon is in its waxing crescent, waxing gibbous, waning gibbous, or waning crescent phase most of the time during any month. Each of the crescent or gibbous phases will last about one week. Figure 42.1 shows the relative positions of Earth, Sun, and moon and the local solar times at which the moon crosses an observer's overhead meridian.

LEARNING OBJECTIVES

After completing this experiment, you should be able to do the following:

1. List the phases of the moon and tell the approximate local solar time when they will be on your overhead meridian.
2. Locate the moon in the sky at various times of day depending upon its phase.
3. Identify the phase of the moon that you are observing.
4. Tell the approximate rising and setting times for the various phases of the moon.

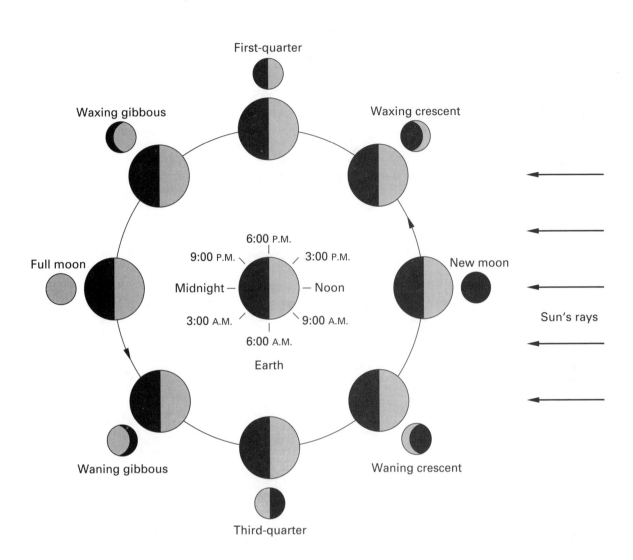

Figure 42.1 *Phases of the moon. The diagram shows the relative positions of the Sun, Earth, and moon. The outer spheres show the moon as seen from Earth's surface. The moon is on the meridian at the local solar times indicated.*

PROCEDURE 1

Use Fig. 42.1 and complete Table 42.1, which can be done anytime after the experiment has been assigned.

PROCEDURE 2

Observe the appearance and position of the moon over a two- to six-week period and record your observations on your personal observation log, Data Table 42.2. Pick a single, convenient location from which to make most of your observations so that you will have a better reference from which to judge the moon's position in the sky. If your schedule does not permit the use of this same location for all observations, you may use other observation points. However you must be able to determine the direction of due south at each location that you use. Your chosen location should satisfy the following conditions:

1. You should not be under bright lights such as street lamps that can affect clear viewing.
2. Most of the sky and the horizon should be visible. The middle of an open space like a field or parking lot is usually suitable.
3. There should be landmarks on the horizon (buildings, trees, and so on), so that your log entries will have a more consistent reference.
4. You must determine the direction of due South at each observing location so that you can establish the location of your overhead meridian.

   Look over the times that the various phases of the moon should be visible and try to see as many of them as possible from your chosen observation point. This means that you will have to look at the sky at several different times, so you may have to make some observations from other locations. Keep a record of your observations over several weeks on your personal observation log, Data Table 42.2. From this table you can confirm the values in Data Table 42.1 that you filled in, showing the rising and setting times of the moon. Remember that the moon is not always in the sky at night. Sometimes you will have to make your observations of the moon while the Sun is above the horizon.

Data Table 42.1   Approximate Local Solar Times of Moonrise and Moonset

Assume a 12-hour period for the moon to be above the horizon. This is, of course, not strictly true for all seasons of the year, but it will give you a rough guide as to when to expect the moon to rise or set in its various phases.

Phase	Moon rises	Moon on overhead meridian	Moon sets
New moon	6:00 A.M.	Noon	6:00 P.M.
Waxing crescent			
First quarter			
Waxing gibbous		9:00 P.M.	
Full moon			
Waning gibbous			
Third quarter			
Waning crescent			

QUESTIONS

1. Why does an observer on Earth always see the same side of the moon's surface?

2. During which phase of the moon can a lunar eclipse occur? During which phase can a solar eclipse occur?

3. Make a drawing illustrating the positions of the Sun, moon, and Earth during a total solar eclipse. Show the shadows of both the moon and Earth. (It is not necessary to make the drawing to scale.)

4. The term *terminator* refers to the boundary line dividing day and night on the surface of a planet or moon. In this experiment, it is the line between the bright and dark side of the moon. State the phase of the moon when the terminator is the (a) sunrise line, (b) sunset line.

Hubble's Law

INTRODUCTION

Most astronomers presently support the theory of an expanding universe. That is, the galaxies or clusters of galaxies that make up the universe are moving away from one another. One reason for believing this theory is the shift in their spectral lines toward the red end (longer wavelengths) of the electromagnetic spectrum. This red-shift is known as the cosmological red-shift.

In 1929 Edwin P. Hubble, American astronomer, published a paper in the Proceedings of the National Academy of Sciences indicating that the radial velocities of some observed galaxies are roughly proportional to their distances. Hubble's discovery, now known as Hubble's law, can be written as

$$v = Hd \tag{43.1}$$

where v = recessional velocity of the galaxy,

d = distance of galaxy from the observer,

H = a constant called the Hubble constant.

The Hubble constant is believed to have a value of 50 to 100 kilometers per second per million parsecs. For example: If H is 50 km/s/mpc, the observed galaxy is moving away from the observer 50 km/s for every one million parsecs (megaparsecs) the galaxy is from the observer. Hubble's law is extremely important because it gives astronomers vital information about the structure of the universe.

LEARNING OBJECTIVES

After completing this experiment you should be able to do the following:

1. State Hubble's law.
2. Determine an experimental value for the Hubble constant.
3. Calculate the approximate age of the universe.

APPARATUS

Hand calculator.

PROCEDURE

Fig. 43.1 is a collection of photographs of three galaxies, their spectra, plus two bright-line spectra for comparing wavelengths. From top to bottom at the left, the photographs show the galaxies at increasing distance from the observer. On the right side are spectra showing H and K calcium lines. They are the two dark vertical lines in the spectrum. The spectrum of each galaxy is bordered, top and bottom, by two bright-line spectra for wavelength reference. The photographs also give the distance to each galaxy in light-years and the recessional velocity in kilometers per second.

In this experiment one objective is to determine the value of Hubble's constant in kilometers per second per megaparsec. To accomplish this, we must convert the distance given in light-years in Fig. 43.1 to megaparsecs. Table 43.1 lists the necessary conversion factors.

Step 1   Data Table 43.1 gives the distances to the three galaxies in light-years. These distances were obtained from Fig. 43.1. Convert these distances to megaparsecs and record in Data Table 43.1.[*] Example:

$$7.8 \times 10^7 \text{ ly} = 7.8 \times 10^7 \text{ ly} \times \frac{1 \text{ pc}}{3.26 \text{ ly}} \times \frac{1 \text{ mpc}}{10^6 \text{ pc}} = 24 \text{ mpc}$$

Step 2   Record the recessional velocities, given in Fig. 43.1, in the data table.

Table 43.1

1 light-year (ly)	= 9.46×10^{12} kilometers (km)
1 parsec (pc)	= 3.26 ly = 3.09×10^{13} km
1 megaparsec (mpc)	= 3.09×10^{19} km
1 day = 86,400 seconds	

Data Table 43.1

	Distance in light-years	Distance in megaparsecs	Recessional velocity km/s	Hubble's constant km/s/mpc
Virgo	7.8×10^7 =			
Corona Borealis	1.4×10^9 =			
Hydra	3.96×10^9 =			
	Average value for Hubble's constant			

_____

[*]See Appendix II for information on how to convert units.

315

CLUSTER NEBULA IN	DISTANCE IN LIGHT-YEARS	RED-SHIFTS

VIRGO — 78,000,000 — H+K — 1,200 KM/SEC

CORONA BOREALIS — 1,400,000,000 — 22,000 KM/SEC

HYDRA — 3,960,000,000 — 61,000 KM/SEC

Figure 43.1 *Relationship between red-shift and distance for extragalactic nebulae. On the left are photographs of three individual elliptical galaxies. From top to bottom the photographs show the galaxies at increasing distance from the observer. On the right, the spectrum (the broad white band) of each galaxy is shown between an upper and lower comparison spectrum. The H and K lines of ionized calcium are the two dark vertical lines in the galaxy's spectrum. The arrows indicate the shift in the calcium H and K lines. The red-shifts are expressed as velocities. (Palomar Observatory Photograph)*

Step 3 Calculate the value of Hubble's constant in kilometers per second per megaparsec for each galaxy and record the value in the data table. Use Eq. 43.1. Determine the average value and record in the data table.

Step 4 Plot a graph with recessional velocity on the y axis and distance on the x axis. Determine the slope of the curve. Refer to Experiment 1 for plotting a graph.

QUESTIONS AND CALCULATIONS

1. How does the slope of the curve compare with the average value of H calculated in Step 3?

2. Determine the value of Hubble's constant in units of kilometers per year per kilometer. To do this, convert units of kilometers per second per megaparsec to kilometers per year per kilometer. Set the average value of Hubble's constant plus units as recorded in Table 43.1 equal to itself (see below). Multiply by one over one, using the correct conversion factors, until the answer is obtained. Show your work in the space below.

$$\underline{\hspace{1cm}} \text{ km/s/mpc} = \underline{\hspace{1cm}} \text{ km/s/mpc} \times \frac{1}{1}$$

3. The age of the universe (a rough estimate) can be obtained from the Hubble constant. The time elapsed since the Big Bang, which is the age of the universe, is the time of separation (distance d) of galaxies receding from one another with a velocity (v). From the equation for velocity ($v = d/t$) we obtain

$$t = d/v$$

from Hubble's law $v = Hd$

substituting

$$t = d/Hd$$

canceling the d's

$$t = 1/H$$

Calculate the value of (t) in years using the value for (H) determined in Question 2. Show your work.

4. How does the calculated value of the age of the universe vary as the value of Hubble's constant increases?

5. How does the force of gravity between galaxies affect the recessional velocity of the galaxies? How would this affect the value of Hubble's constant? The calculated age of the universe? Explain your answers.

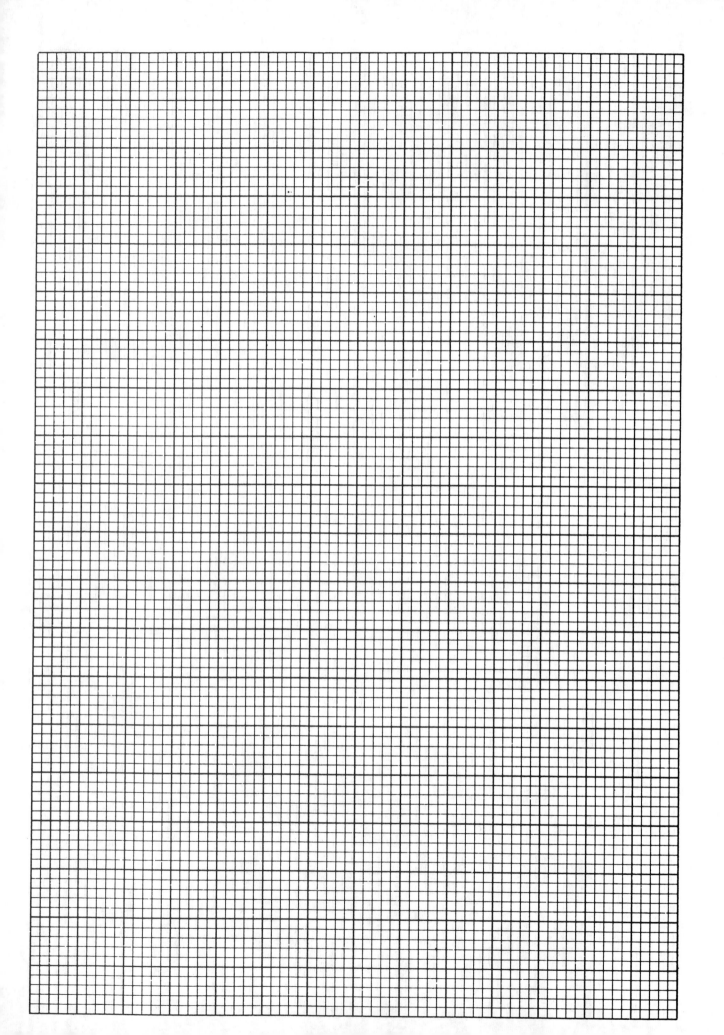

Experiment 44

Air Pressure

INTRODUCTION

Pressure is defined as force per unit area:

$$P = \frac{F}{A}$$

where
- P = pressure,
- F = force in dynes, newtons, or pounds,
- A = area in square centimeters, square meters, or square inches.

The meteorologist uses the millibar as a unit of pressure. The **millibar** is one-thousandth of a bar. The **bar** is defined as one million dynes per square centimeter. The relationships are expressed as follows:

$$1 \text{ millibar} = \frac{1}{1000} \text{ bar}$$

$$1 \text{ bar} = 10^6 \frac{\text{dynes}}{\text{cm}^2}$$

$$1 \text{ millibar} = 10^3 \frac{\text{dynes}}{\text{cm}^2}$$

Air pressure is due to the weight of air. At Earth's surface (45°N, sea level) the total weight of all the air above one square inch of area is 14.7 lb. This is equivalent to a column of mercury 30 in high of one square inch cross-sectional area. This pressure is known also as one atmosphere. When measured in millibars, one atmosphere equals 1013 millibars.

As described by Newton's law of universal gravitation, the density of the atmosphere varies inversely with height above Earth's surface. Density is greater at the surface and decreases with increasing elevation. The density of the atmosphere at sea level is about 1.3×10^{-3} g/cm^3 and decreases to about one-half this value at 18,000 ft.

Swiss physicist Daniel Bernoulli (1700–1782) stated that the properties of gases could be explained with the assumption that any gas consists of minute corpuscles in rapid motion that undergo elastic collisions with one another and with the walls of the container holding them. This bombardment of the walls of the container gives rise to what we call pressure. Bernoulli's statement was made more than 100 years before the kinetic theory of gases was propounded.

Bernoulli also stated a principle about fluids (liquids or gases) in motion. Today this is known as Bernoulli's principle. In simple terms, it states that where the velocity of a fluid is maximum, the pressure is minimum and where the velocity of a fluid is minimum, the pressure is maximum. For example, in Fig. 44.1 the baseball is moving to the right through the air and spinning counterclockwise. This creates a lower pressure on the upper side of the baseball, because the motion of the baseball through the air plus the spin increases air velocity. A higher pressure is created underneath the baseball because the motion of the baseball through the air minus the spin decreases air velocity. With the lower pressure on top and higher pressure underneath, the baseball curves upward.

Figure 44.1 *Spinning baseball curves because of changes in air pressure.*

LEARNING OBJECTIVES

After completing this experiment, you should be able to do the following:

1. Define pressure and state its units of measurement.
2. State the cause of air pressure at Earth's surface and how air pressure varies with altitude.
3. State some of the effects of air at rest.
4. State Bernoulli's principle.
5. State some of the effects produced by air in motion.

APPARATUS

Meter stick, support base and holder to balance meter stick, two toy balloons (round, same size), two rubber bands (small), enamel trays (large white), glass tumbler (round with smooth edge), index card (cover for glass tumbler), two plastic soda straws, straight pin, rubber hose (¼ in in diameter, 2 ft in length), vacuum base, glass plate (4 × 6 in), two Ping-Pong balls, two plastic bottles (250 mL), large beaker (1 L), pinch clamp.

PROCEDURE

1. Determine the least count of the barometer located in the laboratory. _____

2. Determine the present barometric pressure in

   (a) inches of mercury..   _____

   (b) centimeters of mercury...   _____

3. Inflate the two balloons to approximately the same size and with rubber bands secure each near the opposite ends of a meter stick. Balance the meter stick on the support base and then observe the direction the meter stick moves when one of the balloons is punctured with a straight pin. Explain.

4. Place the glass tumbler in the large white enamel tray and fill the tumbler until it overflows. (Use a large beaker for a supply of water from the water tap.) Position an index card over the open end of the glass tumbler to cover the open end completely. Holding one hand firmly on the index card to keep it tight against the rim of the glass, invert the tumbler; then remove your hand from the card. Observe that the water stays in the tumbler. Explain.

5. Close one end of a plastic soda straw with a pinch clamp. You may have to double the end of the straw back slightly before applying the clamp. Puncture the soda straw with a straight pin about 1 in above the clamp. Fill the large enamel tray with water to a depth of about an inch and submerge the straw horizontally in the tray of water. Position your finger over the open end of the straw and raise the straw out of the water to a vertical position with the clamped end downward. Intermittently remove and replace your finger on the open end of the straw. Explain what you observe.

6. Place a large beaker about three-fourths full of water in the large enamel tray, which contains about 1 in water. Lay the rubber hose flat in the tray of water so that it will become filled with water. Holding a finger over each end of the rubber hose so that the water cannot escape, place one end of the rubber hose in the beaker of water and leave the other end in the tray. Remove both fingers from each end of the rubber hose, being careful to keep the end of the hose you have placed in the beaker below the level of the water. Raise the beaker of water and observe what happens. Lift the lower end of the hose above the water level in the tray. This is called a siphon. Explain how it operates.

7. Position the vacuum base in the center of the glass plate and pull the lever forward. Explain why the glass plate is held firmly by the rubber base.

8. Hang two Ping-Pong balls on separate threads so that they are suspended about 20 cm from a support rod and are about 2 cm apart. Using a clean soda straw blow a stream of air between the two balls from a position in front of the balls and level with them. What do you observe? Explain.

9. On a glass plate placed at the edge of the laboratory table, position two plastic bottles on their sides so that they are about 2 cm apart with their long axes parallel to one another. Using a clean soda straw, blow a stream of air between the bottles in a direction parallel to the long dimension of the bottle. Explain what you observe.

QUESTIONS

1. If the atmosphere were condensed to the density of mercury, how high above sea level would the upper level of the liquid be? Use the data from this experiment to answer the question.

2. What is the existing local air pressure in millibars? Show your work for obtaining the answer.

_____mb

3. Is the air pressure in Question 2 considered to be a low or a high pressure? Explain.

4. What pressure units is a meteorologist using when he or she says the barometric pressure is 29.68 and falling?

5. Using information gained from this experiment, explain why

   a. a ship moving in a narrow canal may collide with the side of the canal.

   b. a person in a parked automobile along a freeway feels the car being "pulled" toward a semi-truck as it passes by.

Experiment 45

Humidity

INTRODUCTION

Humidity refers to the amount of moisture or water vapor in the air. The air in our environment is rarely ever dry. The amount of moisture present varies up to about 4% by volume.

Moisture present can be expressed in terms of absolute humidity or relative humidity. **Absolute humidity** is defined as the amount of moisture present in a specified volume of air and is expressed as the actual moisture content (AC). (The moisture is measured in grains. One grain = 1/7000 lb.) In the metric system of units, the cubic meter is normally used to measure volume. In the British system, the volume is measured in cubic feet.

$$AC = \frac{grains}{ft^3}$$

where AC = actual moisture content = absolute humidity (AH).

The **relative humidity** is defined as the ratio of the amount of moisture present in one cubic foot of air (actual moisture content) to the amount of moisture one cubic foot of air can hold (capacity) at the existing air temperature:

$$(RH)_T = \frac{AH}{MC}$$

where $(RH)_T$ = relative humidity at a given temperature,

$\qquad$ AH = absolute humidity,

$\qquad$ MC = maximum moisture capacity at the given temperature.

Relative humidity can be expressed either as a fraction or as a percentage. The fraction can be converted to a percentage by multiplying the fraction by 100:

$$RH = \frac{AH}{MC} \times 100$$

The absolute humidity of an air sample can be determined if the dew point is known. The **dew point** of the air refers to the reading on the thermometer at the time when dew begins to form or the air becomes saturated. If the temperature of the air is gradually lowered, a point (temperature) will be observed at which the air can no longer hold the moisture present and the water vapor condenses to liquid water. Knowing this temperature allows us to determine the amount of moisture present in

327

grains in one cubic foot of air. For example: Refer to Data Table 45.1. If the dew point temperature is 40°F, the absolute humidity is 2.8 grains.

LEARNING OBJECTIVES

After completing this experiment, you should be able to do the following:

1. Define humidity, absolute humidity, and relative humidity, state their units of measurement, and give an example of each.
2. Determine experimentally the absolute and relative humidity of the air.
3. Calculate the relative humidity from given and inferred data.

APPARATUS

Thermometer, small beaker (250 mL), calorimeter cup, ice cubes, small plastic spoon, stirring rod.

PROCEDURE

1. Fill the glass beaker half full of water. Insert the thermometer; then add small pieces of ice to the water, stirring gently with the stirring rod. The temperature of the water and the beaker will begin to lower. Allow the temperature to decrease until you notice moisture beginning to collect on the outside of the beaker. Note the temperature as this occurs and record it in Data Table 45.1.

2. Remove all ice from the glass beaker and allow the water to increase in temperature. Stir gently with the thermometer and determine the temperature point at which the moisture on the outside of the beaker disappears. Record this temperature in the data table.

3. Repeat Procedures 1 and 2, using the calorimeter cup in place of the glass beaker. Record temperatures in Data Table 45.1.

4. Determine the absolute humidity of the air, using Data Table 45.1 and the average value for the dew point. Refer to the example in the introduction. Record the value in Data Table 45.1.

Data Table 45.1

Procedure	Dew point, in degrees Celsius
1.	
2.	
3a.	
3b.	
4. Absolute humidity at room temperature	

CALCULATIONS

1. Determine the average dew point as measured in Procedures 1 and 2.

2. Determine the average dew point as measured in Procedures 3a and 3b.

3. Calculate the relative humidity of the air. Show your work.

QUESTIONS

1. Distinguish between absolute humidity and relative humidity.

2. The air temperature in the laboratory is lowered 5°F. Assuming the absolute humidity remains constant, would the relative humidity increase or decrease? How much? Show your work.

3. If the relative humidity of the air in the laboratory is 25%, what is the absolute humidity?

Table 45.1a   SI Units   Maximum amount of moisture that can be held by one cubic meter of air at various temperatures.

Temperature °C	Moisture kg
0	0.00485
5	0.00341
10	0.00940
15	0.01283
20	0.01730
25	0.02305
30	0.03038
35	0.03963
38	0.04626

Table 45.1b   British Units   Maximum amount of moisture that can be held by one cubic foot of air at various temperatures.

Temperature °F	Moisture in grains
0	0.5
5	0.6
10	0.8
15	1.0
20	1.2
25	1.6
30	1.9
35	2.4
40	2.8
45	3.4
50	4.1
55	4.8
60	5.7
65	6.8
70	8.0
75	9.4
80	10.9
85	12.7
90	14.8
95	17.1
100	19.8

Experiment 46

Weather Maps (Part 1)

INTRODUCTION

Air masses are large bodies of air that take on the characteristics of the area in which they originate. Air masses originating over Canada are cold and dry, whereas air masses originating over the Gulf of Mexico are warm and humid. An air mass is fairly homogeneous in temperature and pressure across its expanse but it shows considerable vertical variation. In general, the pressure and temperature of an air mass decrease with increasing elevation. Because of unequal heat received by Earth's surface from the Sun, the temperature changes; the density of the air is thereby caused to change, and thus the pressure. Such changes coupled with Earth's rotation bring about movement of the air masses over the surface of Earth.

The boundary between two air masses is called a **front**. If a cold air mass is moving into a warm area, the front is labeled a **cold front**. If warm air is moving into a cold region, the front is called a **warm front.**

The term **weather** refers to the day-to-day variations in the general physical properties of the lower part of the atmosphere, known as the troposphere; this region varies from 7 to 11 mi in height. The physical properties that play an active role in our daily lives are temperature, pressure, humidity, precipitation, wind, and extent of cloud formation.

The U.S. National Weather Service is the federal organization that provides national weather information. It is part of the National Oceanic and Atmospheric Administration (NOAA—pronounced "Noah"), which was created within the U.S. Department of Commerce in 1970. Under NOAA, the National Weather Service reports the weather of the United States and its possessions, provides weather forecasts to the general public, issues warnings against tornadoes, hurricanes, floods, and other weather hazards, and records the climate of the United States.

The nerve center of the National Weather Service is the National Meteorological Center (NMC), located just outside Washington, D.C., at Suitland, Maryland. It is the NMC that receives and processes the raw weather data taken at thousands of weather stations. Analyses of data are made twice daily on observations taken over the northern hemisphere at 0000 hours and 1200 hours Greenwich Mean Time (7 A.M. and 7 P.M. EST).

To help analyze the data taken at the various weather stations, maps and charts are prepared that display the weather picture. Fig. 46.1 shows some of the important symbols used to represent the weather in these charts. Since these charts present a synopsis of the weather data, they are referred to as synoptic weather charts. There are a great variety of weather maps with various presentations. However, the most common are the daily weather maps issued by the U.S. National Weather Service. A set consists of a surface weather map, a 500-millibar height contour map, a map of the highest and lowest temperatures, and one showing the precipitation areas and amounts. (See the Daily Weather Maps, weekly series, supplied by the laboratory.)

The Surface Weather map presents station data and analysis for 7 A.M. EST for a particular day. This map shows the major frontal systems and the high- and low-pressure areas at the constant surface level. An **isobar** is a line drawn through points of equal pressure. They are represented by solid lines on the weather map. An **isotherm** is a line drawn through points of equal temperature. Prevalent isotherms are indicated by dashed lines. On the Surface Weather map a station model, shown in Fig. 46.2, gives the pertinent weather data at the station's location. The stations shown on the weather maps in this experiment are only a fraction of those included in the National Weather Service's operational weather maps on which the analyses are based.

LEARNING OBJECTIVES

After completing this experiment, you should be able to do the following:

1. Define air mass, warm front, cold front, isobar, isotherm, and wind.
2. Read weather information from surface weather maps issued by the National Weather Service.
3. Plot station reports similar to those found on Surface Weather maps.

APPARATUS

This *Laboratory Guide* and daily weather maps (weekly series) from the National Weather Service.

Cloudiness

○   ◔   ◑   ◕   ●   ⊗

Amount of cloud cover   Clear   1/4   1/2   3/4   Overcast   Sky obscured

Weather

Rain   •

Snow   ✳

Shower   ▽

Thunderstorm   �R

Freezing rain   ⌒

Fog   ≡

Blowing snow   ⊹

Dust storm or sandstorm   ⌇

The heavier rain or snow is, the more dots or stars are plotted, up to four.

∴• denotes heavy continuous rain.

Solid shading indicates areas where precipitation is currently falling.

Cold front

Warm front

Occluded front

Stationary front

Figure 46.1 *Some of the most important symbols used to display the weather picture.*

SPECIMEN STATION MODEL*

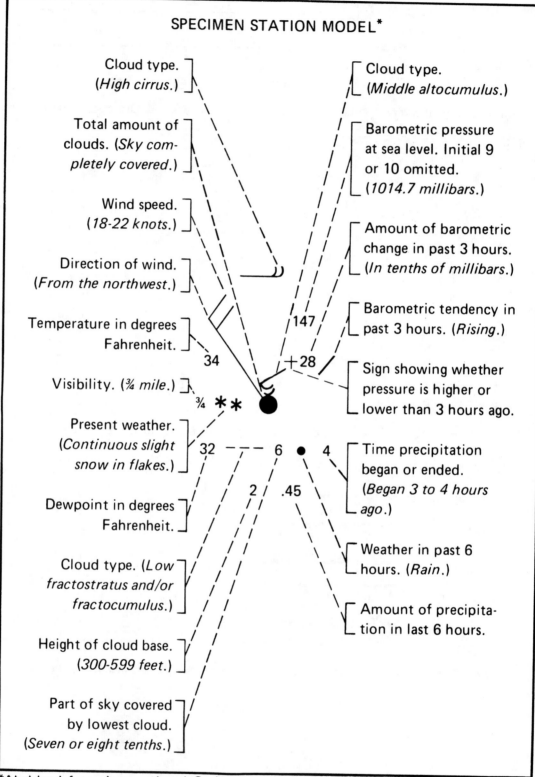

Cloud type. (*High cirrus.*)

Total amount of clouds. (*Sky completely covered.*)

Wind speed. (*18-22 knots.*)

Direction of wind. (*From the northwest.*)

Temperature in degrees Fahrenheit.

Visibility. (*¾ mile.*)

Present weather. (*Continuous slight snow in flakes.*)

Dewpoint in degrees Fahrenheit.

Cloud type. (*Low fractostratus and/or fractocumulus.*)

Height of cloud base. (*300-599 feet.*)

Part of sky covered by lowest cloud. (*Seven or eight tenths.*)

Cloud type. (*Middle altocumulus.*)

Barometric pressure at sea level. Initial 9 or 10 omitted. (*1014.7 millibars.*)

Amount of barometric change in past 3 hours. (*In tenths of millibars.*)

Barometric tendency in past 3 hours. (*Rising.*)

Sign showing whether pressure is higher or lower than 3 hours ago.

Time precipitation began or ended. (*Began 3 to 4 hours ago.*)

Weather in past 6 hours. (*Rain.*)

Amount of precipitation in last 6 hours.

*Abridged from International Code

Figure 46.2 *Specimen station model.*

EXPERIMENT 46   NAME (print) _____ DATE _____

LAST   FIRST

LABORATORY SECTION _____ PARTNER(S) _____

PROCEDURE

1. Record the information called for in Data Table 46.1 for each city listed. The instructor will indicate which weather map is to be used to complete Data Table 46.1.

Data Table 46.1

	**Bismarck, ND	Oklahoma City, OK	Mont-gomery, AL	Houston, TX	New Orleans, LA
Temperature, in degrees F					
Dew point, in degrees F					
Barometric pressure, in millibars					
Wind direction*					
Wind magnitude, in knots					

*A wind is a horizontal movement of air over the surface of Earth. The direction the wind is coming from is the direction of the wind.

**Use these five cities or others announced by the laboratory instructor. If new cities are to be used, cross out listed names and write in the new city names.

2. Plot the information given below for the three cities (A, B, and C) indicated.

	City A	City B	City C
Air temperature	62°F	88°F	32°F
Air dew point	60°F	52°F	26°F
Air pressure	1022.6 mb	987.4 mb	1042.8 mb
Wind direction	SW		NW
Wind magnitude	24 knots	Calm	10 knots
Cloud coverage	0.8	No clouds	0.5
Present weather	Continuous rain		Intermittent fall of snowflakes, moderate

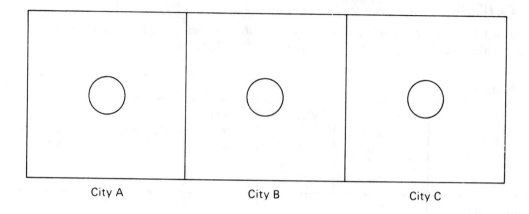

City A   City B   City C

3. Determine the speed of a cold front by observing the front on two consecutive days on the two weather maps in the laboratory manual (Figs. 46.3 and 46.4).

4. Using the weather map supplied by the laboratory instructor, select a cold front and a warm front and record the air temperature that exists about 200 mi on either side of the fronts and indicate whether the temperature rises or falls as the front is crossed at right angles from front to back.

Cold front	_____ °F	_____ °F	_____
Warm front	_____ °F	_____ °F	_____
	Temperature ahead	Temperature behind	Rise or fall

5. Record the barometric pressure that exists about 200 mi on either side of the fronts you selected in Question 4, and indicate whether they rise or fall as the front is crossed at right angles from front to back.

Cold front	_____ mb	_____ mb	_____
Warm front	_____ mb	_____ mb	_____
	Pressure ahead	Pressure behind	Rise or fall

Figure 46.3

Figure 46.4

QUESTIONS

1. List the changes in the temperature and the air pressure that occur as a warm front passes over a surface area.

2. List the changes in the temperature and the air pressure that occur as a cold front passes over a surface area.

3. The Surface Weather map presents data and analysis for what time of day?

4. Define an isobar. How are they represented on the weather map?

5. Define an isotherm. How are they represented on the weather map?

6. How is the amount of precipitation indicated on the weather map?

7. What is the lowest pressure indicated by an isobar on Fig. 46.3?

8. What is the highest pressure indicated by an isobar on Fig. 46.3?

Weather Maps (Part 2)

INTRODUCTION

The most outstanding features shown on a daily weather map (see Experiment 46) are the warm and cold fronts.

In addition to the long black lines with their rounded or sharp-pointed projections that symbolize a warm or cold front, there are plain dark lines with open ends, and others that form complete circles. These lines, called **isobars**, are lines of equal pressure, and their central areas indicate regions of high or low pressure.

A study of the weather map will show many facts concerning the weather elements associated with these pressure cells. The first noticeable feature is the way the isobars are plotted on the map. Note the unit of measurement and the difference between any two isobars drawn on the map. It is clearly seen that reporting stations seldom report these particular values of pressure; the isobars are drawn between the known values indicated at the stations.

Note how the pressure varies as one approaches the center of the high- or low-pressure cell. Also note the fact that no two isobars cross. Why?

LEARNING OBJECTIVES

After completing this experiment, you should be able to do the following:

1. Define high-pressure cell, low-pressure cell, and isobars.
2. Identify and explain the physical characteristics of weather associated with high-pressure and low-pressure cells.
3. Plot pressure values on a map and draw lines of equal pressures (isobars) on the map.

APPARATUS

This *Laboratory Guide* and a weekly series of the daily weather maps from the National Weather Service.

PROCEDURE

1. A high-pressure area is shown over Saskatchewan on both weather maps in Experiment 46. Record the wind direction at approximately 90 ° intervals around the high-pressure cell.

   Wind direction north of central area...   _____

   Wind direction east of central area..   _____

   Wind direction south of central area...   _____

   Wind direction west of central area..   _____

   What is the direction (clockwise or counterclockwise) of the circulation pattern around the cell?

   _____

2. A low-pressure area is shown over eastern Iowa on the weather maps in Experiment 46. Record the wind direction at approximately 90 ° intervals around the low-pressure cell.

   Wind direction north of central area..   _____

   Wind direction east of central area..   _____

   Wind direction south of central area...   _____

   Wind direction west of central area..   _____

   What is the direction (clockwise or counterclockwise) of the circulation pattern around the cell?

   _____

3. Record the maximum pressure indicated on any isobar shown on a weather map in the *Laboratory Guide*. ....   _____

4. Record the minimum pressure indicated on any isobar shown on a weather map in the *Laboratory Guide*. ....   _____

5. Plot the following pressures, given in millibars, on the map of the United States (Fig. 47.1) and draw in the 1020 and the 1024 isobars. See Weather Service map for location of cities.

Estevan, Saskatchewan	1019.4
Winnipeg, Manitoba	1019.8
Armstrong, Ontario	1018.7
Lake Head, Ontario	1018.7
Billings, Montana	1017.9
Miles City, Montana	1019.6
Bismarck, North Dakota	1024.5
Fargo, North Dakota	1024.3
Minot, North Dakota	1020.7
Duluth, Minnesota	1020.7
International Falls, Minnesota	1020.6
Minneapolis, Minnesota	1022.0
Escanaba, Michigan	1015.8
Casper, Wyoming	1018.1
Cheyenne, Wyoming	1020.2

(continued)

Huron, South Dakota	1026.1
Pierre, South Dakota	1025.7
Rapid City, South Dakota	1024.5
La Crosse, Wisconsin	1017.3
Milwaukee, Wisconsin	1014.0
Wausau, Wisconsin	1016.8
Denver, Colorado	1016.8
Pueblo, Colorado	1015.2
North Platte, Nebraska	1023.4
Omaha, Nebraska	1023.7
Valentine, Nebraska	1025.7
Concordia, Kansas	1021.8
Goodland, Kansas	1019.6
Dodge City, Kansas	1018.2
Wichita, Kansas	1018.9
Des Moines, Iowa	1022.4
Sioux City, Iowa	1024.5
Kansas City, Missouri	1020.1
Kirksville, Missouri	1020.1
Springfield, Missouri	1019.1
Chicago, Illinois	1014.1
Moline, Illinois	1017.1
Springfield, Illinois	1016.8

QUESTIONS

1. How is the unit of pressure, the bar, defined? (See Experiment 44.)

2. What is the usual range of atmospheric pressure (in millibars) at Earth's surface?

3. What is the difference in pressure indicated by two adjacent isobars?

4. What is the direction of the air circulation around a low-pressure cell in the northern hemisphere?

5. What kind of weather is usually associated with a low-pressure cell in North America? Why?

6. What is the direction of the air circulation around a high-pressure cell in the northern hemisphere?

7. What kind of weather is usually associated with a high-pressure cell? Why?

8. A pressure of 1012 mb is indicated on a weather map. Is this pressure high or low? Explain.

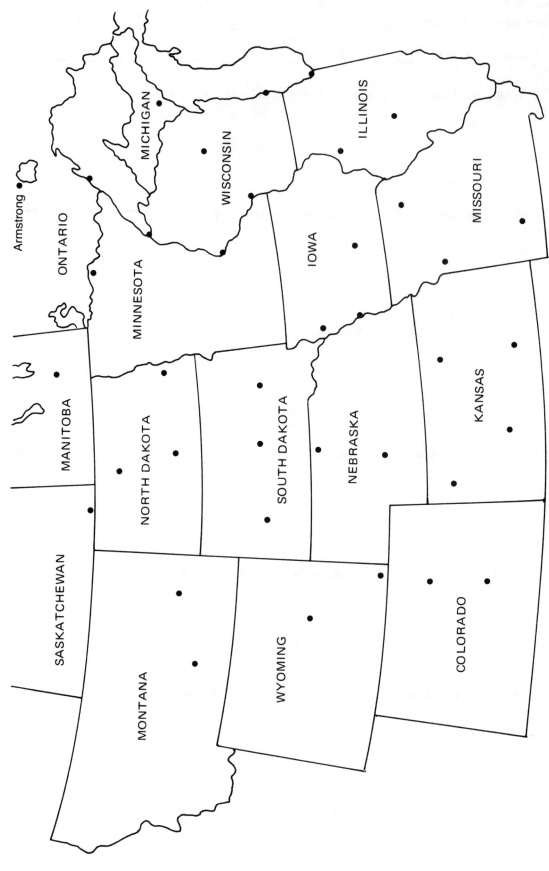

Figure 47.1

Topographic Maps

INTRODUCTION

A **topographic map** is a scaled-down representation of a surface area of Earth as viewed from a high altitude. The maps are generally drawn by means of contour lines plus color and shading to represent specific areas.

The **quadrangle map** issued by the U.S. Geological Survey, Washington, D.C., will be studied in this laboratory experiment.

Direction When a quadrangle map is positioned for normal viewing, geographic north is at the top of the map. The meridian lines traverse the map from bottom to top, and the parallels traverse the map from left to right. The latitude can be obtained by reading the degree values at the side of the map, and longitude can be obtained by reading the degree values at the bottom or at the top of the map.

Colors and Symbols Color, shading, and symbols are used on the quadrangle map to indicate specific items of interest. Green is used to indicate forests. Blue indicates a body of water such as a lake, river, or stream. Salmon color indicates a city or town. Black is used for the boundary lines and artificial structures. Brown is used for the contour lines. Red lines and broken red with white lines indicate primary roads. Secondary roads are drawn as solid faint black or dashed faint black lines. Railroads are indicated as a faint black broken line with black circular dots. Other details are indicated with various shadings of color.

Scale The reduced scale of the map is given as a ratio or fraction, and is usually given at the bottom of the map. A scale of 1:62,500 means that one inch on the quadrangle map represents 62,500 in on Earth's surface. Earth's surface is considered to be horizontal.

Contour Lines Contour lines are drawn in brown and show details of Earth's surface. Any contour line drawn on the map indicates a line of equal elevation. The interval between contour lines is usually 20 ft. This means that there is a 20-ft elevation difference between any two lines. Generally every fifth line is drawn broader and darker than the others. These darker lines are numbered with the correct elevation in feet above mean sea level.

Contour lines that are close together indicate a steep slope; that is, change in elevation takes place rapidly, or the terrain is very steep. See Fig. 48.1 for methods used to draw contour lines to indicate surface profile.

Elevation The elevation of a particular place on the map is indicated in feet above mean sea level. As mentioned above, the dark brown contour lines give the elevation in places that have been determined with care. These are marked BM, meaning bench mark.

The elevation at the bench mark is also indicated in feet. The bench mark is usually a circular brass plate located at the point indicated on the map.

Streams When contour lines are drawn on the map, the direction of a slope can be determined by noting the pattern of certain contour lines. When these lines form a series of V-shaped lines, the point of the V indicates upstream, or to a direction point of higher elevation. Thus, the direction of a stream can be determined. Check this by observing a stream (dotted blue line) on the quadrangle map.

LEARNING OBJECTIVES

After completing this experiment, you should be able to do the following:

1. Read and record data from a quadrangle map.
2. Determine the relationships of groundwater to land surface.
3. Draw a contour map when the surface profile is given.
4. Draw the surface profile represented by a contour map.

APPARATUS

U.S. Geological Survey topographic quadrangles (maps): (1) Local area; (2) Interlachen, Florida.

Figure 48.1 *Surface profile and topographic map (contour interval = 100 ft).*

QUESTIONS

1. Read the introductory material on topographic maps and answer the following questions[*] designated by the instructor.

   (a) What is the map scale?

   (b) What is the total area represented by the quadrangle map? (in square miles)

   ---

   [*]The instructor should change and add questions to fit local quadrangle map.

351

(c) What is the latitude and longitude of the City Hall?

Latitude...

Longitude..

(d) What is the location (direction) of the airport in respect to City Hall?

(e) Draw a contour map to represent the following surface profile.

(f) Draw the surface profile represented by the following contour map.

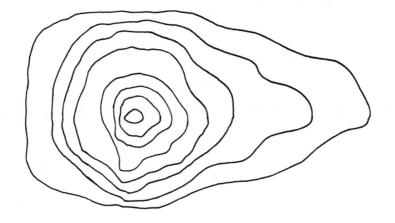

(g) What areas on the map have the least slope?

(h) What is the elevation of _____ Lake?   _____ft
     What is the elevation of _____ Airport?   _____ft
(i) What is the highest elevation shown on the map?   _____ft
(j) State the location of the highest elevation as given in Question (i).

(k) What is the direction of flow of _____ Creek?   _____

2. Interlachen, Florida, quadrangle: Geologically speaking, Florida is our newest mainland state—a "whale's-back" recently risen out of the sea. Thus, the sedimentary rocks are very soft, poorly consolidated coquinas, limy sandstones, and mudstones.

   (a) What is the origin of and the name given to the many depressions (lakes) in the western half of the map?

(b) What is the depth of the depression down to the surface of the lake at Big Pond in the west-central rectangle of the map?

(c) In which direction and at what rate (in feet per mile) does the groundwater table slope or dip in this region?

Use the elevations of Goose Lake (NW rectangle), Fanny Lake (SW), and Twin Lakes (center rectangle) to solve a "three-point" problem: If three different elevations can be obtained from surface outcrops of the same rock bed or from three drill holes, then the slope or dip of any reasonable plane surface can be determined.

(d) Plot the slope of the ground surface at any scale you wish. Now, superimpose on the first slope plane the slope of the water table as determined in the question above. Both profiles should run from the northwestern part of the map to any point within the broad, swampy area in the eastern part of the map.

Do these slope lines explain why the swamps exist where they do? Comment.

Experiment 49

Minerals

INTRODUCTION

A mineral is a solid, homogeneous, inorganic substance (compound or element) found occurring naturally in Earth's crust. Minerals possess a fairly definite chemical composition and a distinctive set of physical properties, which include hardness, cleavage, color, streak, luster, crystalline structure, fracture, tenacity, specific gravity, magnetism, fluorescence, and phosphorescence.

Hardness refers to the ability of one mineral to scratch another. The scale given below is used as a basis for comparing the hardness of some common minerals.

Softest	1.	Talc
↑	2.	Gypsum
		Fingernail
	3.	Calcite
		Copper coin
	4.	Fluorite
	5.	Apatite steel
		Steel knife, plate glass
	6.	Feldspar
		Steel file
	7.	Quartz
↓	8.	Topaz
	9.	Corundum
Hardest	10.	Diamond

Cleavage refers to the tendency of some minerals to break along definite smooth planes. The mineral may exhibit distinct cleavage along one or more planes, or it may exhibit indistinct cleavage or no cleavage. The degree of cleavage that a mineral exhibits is a clue to the identification of the mineral.

Color refers to the property of reflecting light of one or more wavelengths. Although the color of a mineral may be impressive, it is not a reliable property for identifying the mineral, since the presence of small amounts of impurities may cause drastic changes in the color of some minerals.

Streak refers to the color of the powder of the mineral. A mineral may exhibit an appearance of several colors but it will always show the same streak. A mineral rubbed (streaked) across the surface of an unglazed porcelain tile will thereby be powdered and will show its true color.

Luster refers to the appearance of the mineral's surface in reflected light. Mineral surfaces appear to have a metallic or nonmetallic luster. A metallic luster has the appearance of polished metal; a non-metallic appearance may be of varying lusters and the lusters likened to the materials as oppositely listed below:

Adamantine	appearance of a	Diamond
Greasy	appearance of	Oily glass
Pearly	appearance of a	Pearl
Resinous	appearance of	Yellow resins
Silky	appearance of	Silk
Vitreous	appearance of	Glass

Crystalline Structure refers to the way the atoms or molecules that make up the mineral are arranged internally. This arrangement is a function of the size and shape of the molecules and the forces that bind them.

Fracture refers to the way a mineral breaks. The mineral may break into splinters, ragged or rough irregularly surfaced pieces, or shell-shaped forms known as conchoidal fractures.

Tenacity refers to the ability of the mineral to hold together. Some minerals are tough and durable and others are fragile and brittle.

Magnetism refers to the property of possessing a magnetic force field. A mineral possessing magnetism can be detected by a magnetic compass.

Fluorescence refers to the emission of light (to which the eyes are sensitive) by a mineral that is being stimulated by the absorption of ultraviolet or x-ray radiation.

Phosphorescence refers to the emission of light by a mineral after the stimulating source (rays or ultraviolet radiation) has been removed.

LEARNING OBJECTIVES

After completing this experiment, you should be able to do the following:

1. Define a mineral.
2. State the methods and techniques for identifying minerals.
3. Identify some common minerals.

APPARATUS

Mineral specimens, copper coin, plate glass, knife blade or nail, steel file, magnetic compass, magnifying glass, ultraviolet lamp.

Reference for Experiments 43 and 44: W. K. Hamblin and J. D. Howard, *Physical Geology*, Sixth Edition (Minneapolis: Burgess International Group, Inc., 1986).

KEY TO MINERALS

1. Minerals harder than steel or glass (5.5)[*]
    1.1 Minerals with distinct cleavage or with distinct crystal form
        1.11 Minerals with colored streak
            1.111 Pyrite
        1.12 Minerals with uncolored, white, or pale streak
            1.121 Hornblende
            1.122 Potash (or Potassium), Feldspars (Variety: Orthoclase)
                                        (Variety: Microcline)
            1.123 Plagioclase Feldspar (Variety: Albite)
                                 (Variety: Labradorite)
            1.124 Rock Crystal Quartz
    1.2 Minerals with indistinct cleavage or with no cleavage
        1.21 Minerals with colored streak
            1.211 Hematite
            1.212 Magnetite
        1.22 Minerals with uncolored, white, or pale streak
            1.221 Chert
            1.222 Flint
            1.223 Milky Quartz
            1.224 Rose Quartz
            1.225 Jasper
            1.226 Olivine
2. Minerals softer than steel or glass (5.5)
    2.1 Minerals with distinct cleavage or with distinct crystal form
        2.11 Minerals with colored streak
            2.111 Galena
            2.112 Graphite
        2.12 Minerals with uncolored, white, or pale streak
            2.121 Biotite
            2.122 Calcite
            2.123 Gypsum (Variety: Selenite, Satin Spar, Alabaster)
            2.124 Halite
            2.125 Muscovite
            2.126 Azurite
            2.127 Sphalerite
            2.128 Fluorite
            2.129 Talc
    2.2 Minerals with indistinct cleavage or with no cleavage
        2.21 Minerals with uncolored, white, or pale streak
            2.211 Malachite
            2.212 Bauxite

[*]Hardness Scale: 2.5, fingernail; 3.0, copper penny; 5.5, knife blade, window glass; 6.5, steel file.

1.111 **PYRITE**

Hardness: 6–6.5

Streak: black to greenish

Luster: metallic

Color: brass yellow

Chem. Comp.: FeS_2

Sp. Grav.: 5.02

Comments: Often cubic in shape with striations on faces; also comes massive; conchoidal fracture; called "fool's gold."

1.121 **HORNBLENDE**

Hardness: 5–6

Streak: green-gray

Luster: vitreous

Color: greenish black

Chem. Comp.: Ca, Mg, Fe, Al silicate

Sp. Grav.: 3.2

Comments: Long, columnar crystals; visible cleavage in 2 directions; widespread occurrence in igneous and metamorphic rocks, particularly the latter.

1.122 **POTASH (or POTASSIUM) FELDSPARS**

Variety: ORTHOCLASE

Hardness: 6.2

Streak: white

Luster: vitreous

Color: white to gray

Chem. Comp.: $K(AlSi_3O_8)$

Sp. Grav.: 2.57

Comments: Common rock-forming mineral; monoclinic; two prominent cleavages making angle of 90° with each other.

Variety: MICROCLINE

Hardness: 6

Streak: uncolored, white, or pale

Luster: vitreous

Color: buff, green, pink, gray

Chem. Comp.: $K(AlSi_3O_8)$

Sp. Grav.: 2.54–2.57

Comments: Well-developed cleavage in two directions.

1.123 PLAGIOCLASE FELDSPARS

Variety: ALBITE

Hardness: 6

Streak: uncolored, white, or pale

Luster: vitreous

Color: white or light-colored

Chem. Comp.: $Na(AlSi_3O_8)$

Sp. Grav.: 2.62

Comments: Two-directional cleavage.

Variety: LABRADORITE

Hardness: 6

Streak: uncolored, white, or pale

Luster: vitreous

Color: dark color

Chem. Comp.: $Ca(AlSi_3O_8)$

Sp. Grav.: 2.69

Comments: Bluish sheen produced by striations; occasionally shows iridescence.

1.124 ROCK-CRYSTAL QUARTZ

Hardness: 7

Streak: uncolored, white, or pale

Luster: vitreous

Color: colorless

Chem. Comp.: SiO_2

Sp. Grav.: 2.65

Comments: Six-sided crystals with striations on crystal faces; conchoidal fracture.

1.211 HEMATITE

Hardness: 5.5–6.5

Streak: red-brown, except when in powder, then streak is Indian red

Luster: metallic (when in crystals)

Color: black, dark brown, or red

Chem. Comp.: Fe_2O_3

Sp. Grav.: 5.26

Comments: Has a metallic luster when in crystals; also comes in granular, fibrous, or massive forms.

1.212 MAGNETITE

Hardness: 6

Streak: black

Luster: metallic

Color: black

Chem. Comp.: Fe_3O_4

Sp. Grav.: 5.18

Comments: The variety of magnetite called Lodestone is a natural magnet and will strongly deflect a compass needle.

1.221 CHERT

Hardness: 7

Streak: uncolored, white, or pale

Luster: greasy

Color: variable, but usually light

Chem. Comp.: SiO_2

Sp. Grav.: 2.6

Comments: Opaque on thin edge; compact (dense); conchoidal fracture.

1.222 FLINT

Hardness: 7

Streak: uncolored, white, or pale

Luster: greasy

Color: variable, but usually dark

Chem. Comp.: SiO_2

Sp. Grav.: 2.6

Comments: Compact (dense); conchoidal fracture; translucent on thin edge.

1.223 MILKY QUARTZ

Hardness: 7

Streak: uncolored, white, or pale

Luster: vitreous or greasy

Color: milky

Chem. Comp.: SiO_2

Sp. Grav.: 2.65

Comments: Translucent; no cleavage; milky color due to minute liquid inclusions.

1.224   ROSE QUARTZ

Hardness: 7

Streak: uncolored, white, or pale

Luster: vitreous or greasy

Color: pale pink to rose red

Chem. Comp.: SiO_2

Sp. Grav.: 2.65

Comments: Color due to minute traces of titanium; occasionally coarsely crystalline but usually without crystal form.

1.225   JASPER

Hardness: 7

Streak: pale

Luster: vitreous

Color: red to brown

Chem. Comp.: SiO_2

Sp. Grav.: 2.65

Comments: Red color caused by inclusion of hematite; conchoidal fracture. It is a cryptocrystalline quartz (submicroscopic crystals).

1.226   OLIVINE

Hardness: 6.5–7

Streak: uncolored, white, or pale

Luster: vitreous

Color: olive to gray green

Chem. Comp.: $(Mg, Fe)_2SiO_4$

Sp. Grav.: 3.27–4.37

Comments: Granular appearance in mass; transparent variety known as peridot; fracture is visible in peridot; conchoidal.

2.110   GALENA

Hardness: 2.5

Streak: gray or black

Luster: metallic

Color: lead-gray

Chem. Comp.: PbS

Sp. Grav.: 7.4–7.6

Comments: Opaque, heavy, often in cubes, cubic cleavage in three directions at right angles.

2.112   **GRAPHITE**

Hardness: 1

Streak: gray or black

Luster: greasy

Color: gray

Chem. Comp.: C

Sp. Grav.: 2

Comments: Has greasy feel; perfect cleavage in one direction.

2.121   **BIOTITE (BLACK MICA)**

Hardness: 2.5–3

Streak: uncolored, white, or pale

Luster: vitreous

Color: dark green to black

Chem. Comp.: (K, Mg, Fe, Al) silicate

Sp. Grav.: 2.8–3.2

Comments: Basal cleavage prominent (perfect in one direction), flakes in flat sheets; flexible and elastic; transparent to translucent.

2.122   **CALCITE**

Hardness: 3

Streak: uncolored, white, or pale

Luster: vitreous

Color: sometimes colorless but usually white, though it may be any color

Chem. Comp.: $CaCO_3$

Sp. Grav.: 2.71

Comments: Transparent to translucent, rhombohedral cleavage (three inclined directions *not* at right angles); Iceland Spar variety: clear and double refractive. (Make pencil dot on paper, place calcite over it, and dot appears double.)

2.123   **GYPSUM**

Hardness: 2

Streak: uncolored, white, or pale

Luster: vitreous (Satin Spar variety-silky luster)

Color: white or light-colored

Chem. Comp.: $CaSO_4 \cdot 2H_2O$

Sp. Grav.: 2.32

362

Comments:

Variety: Selenite–often clear and transparent; flexible but inelastic sheets; cleavage in three directions, one very good, the others poorer.

Variety: Satin Spar–silky luster, fibrous-forming veinlets.

Variety: Alabaster–composed of aggregates of fine crystals.

2.124 HALITE

Hardness: 2.5

Streak: uncolored, white, or pale

Luster: vitreous

Color: colorless to white

Chem. Comp.: NaCl

Sp. Grav.: 2.16

Comments: Salty taste; water soluble; transparent to translucent; cubic cleavage (perfect in three directions at right angles).

2.125 MUSCOVITE (LIGHT-COLORED MICA)

Hardness: 2–2.5

Streak: uncolored, white, or pale

Luster: vitreous

Color: white or light shades of green or yellow

Chem. Comp.: $KAl_2(AlSi_3O_{10})(OH)_2$

Sp. Grav.: 2.76–3.1

Comments: Basal cleavage prominent (perfect in one direction), flakes in flat sheets; flexible and elastic; transparent to translucent.

2.126 AZURITE

Hardness: 1–3

Streak: uncolored, white, or pale

Luster: vitreous

Color: azure blue

Chem. Comp.: $Cu_3(CO_3)_2(OH)_2$

Sp. Grav.: 3.77

Comments: Transparent to translucent when in crystal form.

2.127 SPHALERITE

Hardness: 3.5–4

Streak: yellow brown to nearly white

Luster: resinous

Color: black to colorless (colorless rare)

Chem. Comp.: ZnS

Sp Grav.: 3.9–4.1

Comments: Often shows distinct cleavage planes.

2.128   FLUORITE

Hardness: 4

Streak: uncolored, white, or pale

Luster: vitreous

Color: can be colorless, yellow, blue, green, or violet

Chem. Comp.: CaF_2

Sp. Grav.: 3.18

Comments: Transparent to translucent, cubic crystals, good cleavage in four directions.

2.129   TALC

Hardness: 1

Streak: uncolored, white, or pale

Luster: pearly or greasy

Color: white, gray-green, silver-white

Chem. Comp.: $Mg_3(Si_4O_{10})(OH)_2$

Sp. Grav.: 2.7–2.8

Comments: Very soft; greasy or soapy feel, foliated or compact masses; perfect cleavage in one direction, translucent on thin edge.

2.211   MALACHITE

Hardness: 3.5-4

Streak: uncolored, white, or pale

Luster: vitreous (when in crystals)

Color: bright green

Chem. Comp.: $Cu_2CO_3(OH)_2$

Sp. Grav.: 3.9–4.03

Comments: Silky in fibrous varieties, dull in earthy type.

2.212   BAUXITE

Hardness: 1–3

Streak: uncolored, white, or pale

Luster: earthy

Color: gray to buff or brown

Chem. Comp.: mixture of aluminum hydroxides

Sp. Grav.: 2.0–2.55

Comments: An earthy rock, composed chiefly of gibbsite, boehmite, and diaspore.

PROCEDURE

Using the mineral specimens provided and referring to the Key to Minerals and Mineral Descriptions on the previous pages, examine each specimen, record its identifying physical properties, determine its position in the Key, name the mineral, and record it in Data Table 49.1. The laboratory instructor will assist you, if necessary.

Data Table 49.1

Specimen number	Mineral	Cleavage	Hardness	Color	Streak	Luster	Tenacity	Magnetism
1								
2								
3								
4								
5								
6								
7								
8								
9								
10								
11								
12								
13								
14								
15								
16								
17								
18								
19								
20								

Rocks

INTRODUCTION

A rock is an independent geologic unit of Earth's crust. Most rocks are composed of a single mineral or combination of two or more minerals, but some are composed of organic materials (for example, bituminous coal). Rocks are placed in three main classes according to their origin:

1. **Igneous rocks**, formed from molten material called **magma**.
2. **Sedimentary rocks**, formed by the accumulation of sediments caused by weathering.
3. **Metamorphic rocks**, formed by a change in the structure of igneous and sedimentary rocks through the application of high pressure and temperature.

Rocks are identified on the basis of texture (coarse grain, fine grain, or glassy) and color, which is a function of the mineral content. The identification of rocks in this experiment will be done with charts that give the description of some common rocks.

LEARNING OBJECTIVES

After completing this experiment, you should be able to do the following:

1. Define the term rock, and state the three classifications of rocks.
2. Identify some common rocks.

APPARATUS

Numbered rock specimens and magnifying glass.

IGNEOUS ROCKS

1. Light-colored rocks
    1.1 Coarse grain, visible to the eye; grains even-sized
        1.11 Quartz present — *Granite*
        1.12 No quartz present — *Syenite*
    1.2 Fine grained — *Rhyolite*
    1.3 Glassy; no graining visible; glassy luster visible under magnifying lens
        1.31 Smooth texture; color light; light in weight; vesicular — *Pumice*

2. Dark-colored rocks
    2.1 Coarse grain, visible to the eye; grains even-sized
        2.11 Hornblende present — *Diorite*
        2.12 Olivine present — *Olivine Gabbro*
    2.2 Fine-grained; grains even-sized
        2.21 Nonvesicular — *Basalt*
        2.22 Vesicular; glassy under lens — *Scoria*
    2.3 Glassy; no graining visible; smooth texture, nonvesicular; conchoidal fracture — *Obsidian*

METAMORPHIC ROCKS

1. Coarse foliation; banded, even-grained, grains fine to coarse; contains feldspars, quartz, and biotite mica — *Gneiss*

2. Fine foliation, even- or uneven-grained, grains medium to coarse
    2.1 Essential minerals mica and quartz — *Mica Schist*
    2.2 Essential minerals mica and quartz, accessory mineral garnet — *Garnetiferous Mica Schist*
    2.3 Green color, silky texture — *Chlorite Schist*

3. Slaty cleavage; grains even in size and very fine-grained — *Slate*

4. Massive; little or no foliation; even-grained, fine to coarse grains
    4.1 Essential mineral calcite or dolomite — *Marble*
    4.2 Essential mineral quartz — *Quartzite*

SEDIMENTARY ROCKS

1. Typically coarse, *angular* fragments cemented in finer materials, vary from pebble size upward — *Breccia*

2. Typically coarse, rounded fragments cemented in finer materials, vary from pebble size upward — *Conglomerate*

3. Sand grains, cemented together by quartz, calcite, or iron oxides; rounded quartz grains most abundant constituent. Granular; medium to fine grains (size of sugar grains)
   3.1 Iron oxide cement
      3.11 Red color — *Red Sandstone*
      3.12 Brown; contains abundant mica, usually muscovite — *Micaceous Sandstone*

4. Clay and silt particles; grains smaller than sand and barely visible. Well compacted; clayey odor.
   4.1 Red, brown, or yellow color — *Ferruginous Shale*

5. Altered plant remains; black, massive, unlaminated; grains indistinguishable; smooth, dull glassy luster; does not soil hands and paper — *Cannel Coal*

6. Variable texture, crystalline, granular, coarse or fine; effervesces in hydrochloric acid
   6.1 Very fine-grained; compact, dense; conchoidal fracture — *Lithographic Limestone*
   6.2 Composed largely of shells — *Shell Limestone*
   6.3 Composed of small masses about the size of a pinhead or smaller — *Oolitic Limestone*

PROCEDURE 1

Complete the information for the introduction to igneous, sedimentary, and metamorphic rocks in Data Table 50.1.

Data Table 50.1   Introduction to Igneous, Sedimentary, and Metamorphic Rocks

Examine the numbered specimens and decide on their distinctive features, keeping in mind that some of the features may also be visible in other types or groups and that some may not appear in other specimens of the same type or group. The following questions should be answered concerning properties of the rock name and rock group given.

1. Rock Name **Obsidian**   Rock Group **Igneous**

   a. Is this rock intrusive or extrusive by formation?   (Circle one.)
   b. What evidence is there to substantiate your answer?

   _____

   c. What do you suppose the general mineral composition of the rock might be?

   _____

2. Rock Name **Granite**   Rock Group **Igneous**

   a. Compared to obsidian, did this rock cool rapidly or did it cool at a slower rate? (Circle one.)
   b. What evidence do you find in this rock to substantiate your answer?

   _____

   c. What are two essential minerals that compose this specimen?

   (1)_____ (2)_____

   d. What is the black accessory mineral found in the specimen? _____

3. Rock Name **Micaceous Sandstone**   Rock Group **Sedimentary**

   a. What is the composition of the lighter colored grains?

   _____

   b. Are they equal-granular (well sorted)? _____

   c. What rock sources might these grains be derived from?

   (1) _____ (2) _____

4. Rock Name **Chlorite Schist** Rock Group **Metamorphic**

    a. This rock has a foliated texture. How is this texture formed?

    _____

5. Rock Name **Marble** Rock Group **Metamorphic**

    a. What is the relative hardness of this rock sample? _____

    b. What is the chemical composition of this rock sample? _____

    c. This rock has undergone metamorphism. What do you think the original rock might have been?

    _____

6. Rock Name **Lithographic Limestone** Rock Group **Sedimentary**

    a. What is the relative hardness of this rock sample? _____

    b. What is the chemical composition of this rock sample? _____

    c. What may be a good interpretation for the environment of the deposition for this sample (fluvial, marine, aeolian, continental)? (Circle one.)

7. Rock Name **Scoria** Rock Group **Igneous**

    a. Is this rock intrusively or extrusively formed?   (Circle one.)
    b. What evidence is there to substantiate your answer?

    _____

8. Rock Name **Sandstone, red** Rock Group **Sedimentary**

    a. What causes the red coloring in this sample? _____

    b. What is the mineral composition of the grains in this sample?

    _____

9. Rock Name **Pumice** Rock Group **Igneous**

    a. How does this sample differ from sample No. 7? _____

    _____

    b. Is this sample intrusively or extrusively formed?   (Circle one.)

10. Rock Name **Shell Limestone** Rock Group **Sedimentary**

    a. What types of fossils are found in this sample? What part of the ocean did they probably come from (beach, reef, deep ocean)?

    _____

    b. What is the chemical composition of this rock sample? _____

11. Rock Name **Quartzite**   Rock Group **Metamorphic**

    a. This is a granular metamorphic rock consisting essentially of quartz. What do you think the original rock was before metamorphism changed it?

    _____

12. Rock Name **Slate, gray**   Rock Group **Metamorphic**

    a. This is a foliated rock. What general rock type could it have been before the effects of metamorphism?

    _____

    b. What is the economic use of this rock? _____

13. Rock Name **Basalt**   Rock Group **Igneous**

    a. What is the probable composition of this rock? _____

    b. Is it intrusively or extrusively formed? (Circle one.)

14. Rock Name **Gneiss**   Rock Group **Metamorphic**

    a. Explain the parallel alignment of the individual layers.

    _____

    b. Which has undergone a higher degree of metamorphism, Rock No. 12 or Rock No. 14? (Circle one.)

15. Rock Name **Cannel Coal**   Rock Group **Sedimentary**

    a. This is a variety of bituminous coal of uniform and compact fine-grained texture with general absence of banded structure; composed predominantly of plant spores.

PROCEDURE 2

Identify each of the specimens provided, and record in Data Table 50.2.

Data Table 50.2

Specimen number	Rock name	Texture	Composition mineral content
1			
2			
3			
4			
5			
6			
7			
8			
9			
10			
11			
12			
13			
14			
15			
16			
17			
18			
19			
20			

Rock-Forming Minerals

INTRODUCTION

Of the more than one hundred chemical elements, only eight (oxygen, silicon, aluminum, magnesium, iron, calcium, sodium, and potassium) combine in different proportions and manner to make up most of the common minerals. Of the two thousand or so recognized species of minerals, only a score comprise by far the greater amount of the common rocks observed on Earth's surface. These few prevalent minerals can be termed the **rock-forming minerals**.

Most igneous rocks consist of **crystals** of two or more of the following minerals: olivine, pyroxene, amphibole, plagioclase feldspar, potassium feldspar (orthoclase), biotite, muscovite, and quartz.

Some sedimentary rocks, notably the evaporites, also consist of crystals, but very often of only a single mineral. For example, rock gypsum is made up of the mineral gypsum, and rock salt is largely the mineral halite; some limestones are almost pure calcite. On the other hand, most sedimentary rocks contain more or less rounded particles or grains of several minerals cemented together. Although a few standstones may have over 90% quartz, most sandstones are generally mixtures of clay minerals and several of the minerals named above; that is, the rock is the expected residue of weathered igneous rocks.

Most metamorphic rocks also contain several of the rock-forming minerals, but in addition they usually have suites of less common minerals that were created by pressures, heat, and solutions acting upon and altering the original materials during the metamorphic processes.

LEARNING OBJECTIVES

After completing this experiment, you should be able to do the following:

1. Name the principal rock-forming minerals.
2. From supplied minerals, determine the bonding, or consolidation mechanism of igneous rocks, as contrasted with that of sedimentary rocks.

APPARATUS

Mineral specimens, plate glass, knife blade, steel file, magnifying glass, key to mineral identification.

PROCEDURE

1. From the minerals in the study tray, identify by number the igneous rock-forming minerals listed below. The instructor will have a set of labeled minerals so that you may later check your identifications.

Data Table 51.1

Specimen number	Mineral	Identifying remarks
	Orthoclase feldspar	
	Plagioclase feldspar	
	Olivine	
	Quartz	
	Amphibole/Pyroxene	
	Biotite	
	Muscovite	
	Pyrite	
	Magnetite	

2. The instructor will provide a specimen of coarsely crystalline granite.

   (a) List the minerals you can identify in the granite.

   (b) Which minerals show cleavage?

   Which important mineral has no cleavage?

   (c) Examine the granite specimen with the magnifying glass or under the microscope. Are there any rounded grains?

   Which minerals show the best crystal shapes?

   Which single mineral shows the greatest range in size and the poorest geometrical shape?

Igneous rocks that form from magma slowly cool and crystallize. Which mineral appears to have crystallized early?

Which mineral seems to have formed quite late?

What is your reasoning for the answer to the above question?

3. (a) How are granites and other igneous rocks bonded or held together? As an aid in answering this question, examine under the microscope the thin section provided by the instructor.

(b) As a contrast, examine the thin section of sedimentary rock (in this case a "clean" sandstone) under the microscope. How do the mineral grains in this rock differ in shape from those in any igneous rock? Of what mineral is this rock almost wholly composed? (See the hand specimen also.)

How are sedimentary rocks, in general, consolidated or held together? (Ask the instructor for a hint, if necessary.)

Igneous Rocks and Crystallization

INTRODUCTION

As mentioned in Experiment 51, most igneous rocks are composed of two or more minerals that are silicates. Experiments done in the laboratory show that the crystallization of silicate materials from molten magma occurs within a temperature range of 1200° to 600° Celsius. The minerals that crystallize out at the higher temperatures show well-defined crystal forms. This is to be expected, because they have greater freedom to grow within the molten melt. Minerals forming at the lower temperatures will develop with restricted freedom and the crystals thus formed will not have well-formed faces.

It is not a difficult task to let a mineral cool and note the crystallization, but it should be remembered that the actual conditions within a cooling magma are complex, and the resulting minerals formed by crystallization depend on many factors.

LEARNING OBJECTIVES

After completing this experiment, you should be able to do the following:

1. Describe crystallization of a warm melt, which is analogous to the crystallization of igneous rocks from a magma.
2. Identify a few igneous rocks and ascertain their textures.

APPARATUS

Igneous rock specimens, glass slides and/or evaporation dishes, forceps, liquid dropper or wire loops, thymol crystals, saturated solutions of several salts. *Note:* thymol may irritate the eyes or skin; handle its solution with the dropper and its crystals with the forceps.

PROCEDURE

1. On the front table are vials of saturated solutions of sodium chloride, sodium nitrate, potassium aluminum sulphate (alum), and copper acetate. Place two or three drops of several of the solutions on a glass slide or in an evaporation dish. Label each specimen, allow to evaporate slowly in the Sun or over a hot plate set on very low heat.

2. From the melted thymol that the instructor provides, place a large drop near the end of a glass slide and allow it to spread to the size of a dime. Using forceps, drop two or three crystals of thymol at scattered points into the melted thymol and observe quickly with the magnifying glass or under the microscope.

   (a) Do the crystals grow by internal expansion or by external accretion?

   (b) Do the crystals show growth lines?

   (c) What is the effect of limited space upon the shapes of some crystals?

3. Place several crystals of thymol near the end of a glass slide and crush them to powder with another glass slide or other smooth object. Using a dropper or wire with a tiny loop, place a drop of melted thymol on the powdered thymol. Examine the crystallization as before.

   (a) How does the crystallization in this experiment differ from that in the above?

   (b) Explain the difference in terms of the spontaneity of crystallization. The same result can be produced by placing melted thymol on a glass slide and then immediately cooling it on an ice cube for some seconds.

   (c) What effect, then, does the rate of cooling have upon crystal size?

4. Although all igneous rocks crystallize from a cooling silicate melt (a magma), some sedimentary rocks, the evaporites, crystallize by the evaporation of cool saline solutions. Examine the specimens you prepared previously in Procedure 1. Note that, for any particular crystalline substance, the angle between corresponding crystal faces is always the same.

   (a) Does each compound have a unique crystal shape?

   (b) Do you think that some minerals may be identified by crystal shape alone?

5. Name each of the numbered igneous features in Fig. 52.1.

   #1 _____   Name any two features that usually have coarsely crystalline textures. _____

   #2 _____

   #3 _____   _____

   #4 _____   Suggest one feature that might show porphyritic texture.

   #5 _____

   #6 _____   _____

   #7 _____   Where would one expect to find rocks showing glassy texture?

   #8 _____   _____

   Which igneous body is older, #5 or #7? Why?

6. Refer to Fig. 52.2 and identify the igneous rocks provided in the specimen tray. In the spaces in Data Table 52.1, briefly describe the texture of each rock.

Data Table 52.1

Rock	Number	Texture
Granite		
Basalt		
Obsidian		
Andesite		
Scoria		

Figure 52.1 *Diagram of igneous features.*

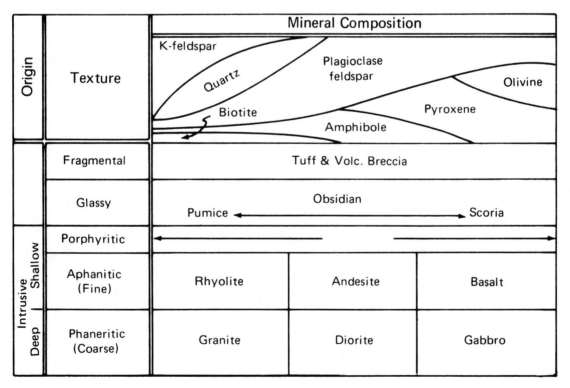

Origin	Texture	Mineral Composition		
		K-feldspar / Quartz / Biotite	Plagioclase feldspar / Amphibole	Olivine / Pyroxene
	Fragmental	Tuff & Volc. Breccia		
	Glassy	Pumice ← Obsidian → Scoria		
Shallow	Porphyritic	← ────── ────── →		
	Aphanitic (Fine)	Rhyolite	Andesite	Basalt
Deep	Phaneritic (Coarse)	Granite	Diorite	Gabbro

Figure 52.2 *An aid in rock identification.*

Sedimentary Rocks

INTRODUCTION

Four processes are involved in the creation of sedimentary rocks: (a) physical and chemical weathering of the parent rock; (b) transportation of the weathered products by gravity, running water, wind, or ice; (c) deposition, with some sorting according to sizes of grains, in sedimentary basins such as lakes, stream channels, and especially the continental shelves of the oceans; (d) compaction and cementation of the sediment into solid rock.

Sedimentary rocks are composed largely of one or more of only four materials: quartz, clay, rock fragments, and shells composed of calcium carbonate (calcite). Most sedimentary rocks are termed **clastics**; that is, they consist of "broken" particles weathered from preexisting rocks. The second most important group of sedimentary rocks are of biochemical and chemical origin, and some, like rock salt and gypsum, are crystalline in texture, having been formed by the evaporation of saline lake and ocean waters. Rarely, organic matter alone forms rock, such as coal. The three groups of sedimentary rocks are outlined in Table 53.1.

LEARNING OBJECTIVES

After completing this experiment, you should be able to do the following:

1. Describe several properties of clastic sedimentary rocks.
2. Identify some common sedimentary rocks.

APPARATUS

Coarse sand, mixture of sands, a glass jar, a graduated cylinder, a millimeter scale, tray of rock specimens, dilute hydrochloric acid, dropper.

Table 53.1

A. CLASTIC

Texture	Composition	Rock name
Coarse-grained (over 2 mm)	Rounded fragments of any rock type, usually quartz, quartzite or chert	Conglomerate
	Angular fragments of any rock type	Breccia
Medium-grained (1/16 mm to 2 mm)	Over 90% quartz	Quartz sandstone
	Quartz with 25% feldspar	Arkose
	Quartz, clay, and rock chips	Graywacke
Fine-grained (1/256 mm to 1/16 mm)	Quartz and clay	Siltstone
Very fine-grained (less than 1/256 mm)	Clay and quartz	Shale

B. BIOCHEMICAL and CHEMICAL

Texture	Composition	Rock name
Medium-to-coarse crystalline	Calcite ($CaCO_3$)	Crystalline limestone
Very fine crystalline		Lithographic limestone
Composed of tiny pellets with concentric internal structure		Oolitic limestone
Fragmented shells loosely cemented		Coquina
Abundant shells in dense calcareous matrix		Fossiliferous limestone
Microscopic shells and clay		Chalk
Banded calcite, often spongy		Travertine
Crystalline replacement of limestones	Dolomite ($CaMg(CO_3)_2$)	Dolomite
Microcrystalline	Chalcedony (SiO_2)	Chert
Fine-to-coarse crystalline	Gypsum $CaSO_4 \cdot 2H_2O$	Gypsum
Fine-to-coarse crystalline	Halite (NaCl)	Rock salt

C. ORGANIC

Texture	Composition	Rock name
Dense, banded	Carbon matter and silt	Coal

PROCEDURE

1. Measure out on a piece of paper enough of the coarse sand at the front desk to make a column about 1 in high in the glass jar or graduated cylinder. Drop or quickly pour all of the sand into the jar or cylinder without agitating the vessel. Measure carefully the height of the top of the sand above the table top.

   (a)   Height of sand above the table... _____ mm

   Now tamp the base of the jar or cylinder lightly half a dozen times on the palm of your hand and remeasure the height of the sand top.

   (b)   Height of sand after tamping.. _____ mm

   All newly deposited sediments eventually compact and become more dense because of the weight of other sediments being deposited on top.

   (c)   By what proportion or percentage did the sample compact? Percentage = 100 *
         Decrease/Original .... _____

   Ignore the thickness of the bottom of the container. Leave the sample in the container to be used in the next procedure.

2. Calculate in cubic centimeters the overall volume of the cylindrical sample used above. If the sample is in a graduated cylinder, volume can be read directly; one millimeter equals approximately one cubic centimeter.

   Volume = 3.14 (R^2H), where R is radius and H is height of column after tamping.

   (a)   Bulk volume of sample, V_b.. _____ mL

   Noting carefully the amount of water used, very slowly add water to the sample so that it comes just to the top of the sand.

   (b)   Volume of water added, V_w .... _____ mL

   (c)   Calculate the porosity or void space of the sand sample:

   $$\text{Percent porosity} = \frac{V_w}{V_b} \times 100 \qquad _____ \%$$

3. Obtain a few spoonfuls of the mixed sands at the front desk. Fill a jar or cylinder somewhat more than half-full of water. Add the mixed sand and check to see if different beds or layers of sediment have formed on the bottom. If not, shake and invert the jar and then allow the sediment to resettle to the bottom.

   (a) Observe the two different layers. What properties of the sediments tell you that there are indeed separate layers present?

(b) If sand and silt were being transported by a stream, which would be moved along the bottom and which carried by suspension? .... _____

(c) If sand and silt were being delivered simultaneously to the sea, which would be deposited first?

4. The tray on the laboratory table contains several sedimentary rocks. Use Table 53.1 and the information on sedimentary rocks to identify the rocks listed in Data Table 53.1.

Data Table 53.1

Number	Rock name	Texture	Mineral composition
	Quartz sandstone		
	Fossiliferous limestone		
	Shale		
	Gypsum or Rock salt		
	Arkose (sandstone)		
	Coquina		
	Oolitic limestone		
	Conglomerate		

Measurement and Significant Figures

A **measurement** is a comparison of an unknown quantity with a precisely specified quantity called a standard unit. Measurements should be made and recorded as accurately as possible. The term **accuracy** refers to how close the measurement comes to the exact value. Accuracy depends upon the precision of the measuring instrument, and the ability of the individual taking the measurement. Errors will be made in every measurement, but the magnitude of the errors can be kept small when the observer is careful and uses an instrument with high precision. **Precision** refers to the degree of reproducibility of a measurement; that is, to the maximum possible error of the measurement. Precision may be expressed as a plus or minus correction. All measurements are approximate. It is impossible to know the "exact" length, mass, or amount of anything. The observer and the measuring instrument place a limit on the accuracy of all measurements.

Instruments for taking measurements are constructed with a calibrated scale for obtaining numerical values concerning the property being measured. The smallest division on the calibrated scale that can be read by the observer, without guessing, is known as the **least count** of the instrument. For example: A meter stick is divided into 100 equal divisions and marked on the stick as centimeters. Each centimeter is further divided into 10 equal divisions. Thus, the least count of the meter stick is one-tenth of one centimeter, or one-thousandth of one meter.

When the meter stick is used to take a measurement of an unknown quantity (for example, length), the observer can always obtain a value within one-thousandth (0.001) of a meter of the exact value of the unknown length, without guessing. But an additional step can increase the accuracy by one doubtful digit, because the observer can estimate a fractional part of the smallest division on the meter stick. For example, if one end of the meter stick is placed at one end of the object to be measured, and the other end of the object falls between two of the smallest divisions, the observer estimates this additional value and adds it to the known scale reading. This estimated digit is significant and is the last digit recorded when taking a measurement.

When the digits are recorded, the number recorded contains all known digits of the measurement plus the doubtful digit. This recorded number is a significant figure. By definition, a **significant figure** is a number that contains all known digits plus one doubtful digit.

Recorded numbers may contain the digit zero (0), which may or may not be significant. When a zero digit is used to locate the decimal point, it is not significant. For example, the numbers 0.048, 0.0032, and 0.00057 each have two significant digits. When a zero appears between two nonzero digits in a number, it is significant. For example, 2.04 has three significant digits; 8.002 has four significant digits. Zeros appearing at the end of a number may or may not be significant. For example, in the number 5480 if the digit eight is a doubtful digit, then the zero is not significant. If the eight is a known digit and the zero is a doubtful digit, then the zero is significant.

The powers-of-10 notation can be used to remove the ambiguity concerning a number like 5480. When using the powers-of-10 notation, we first write all the significant digits of the number. This is followed by 10 to the correct power to locate the decimal point. For example, for three significant digits we write 5.48×10^3; for four significant digits we write 5.480×10^3.

The following procedure is usually used to round off significant figures to fewer digits. If the last significant digit on the right is less than 5, drop it and insert zero instead. If the last significant digit on the right is 5 or greater, drop it and increase the preceding digit by one.

Examples

Round off the following numbers to two significant digits.

247   Because the last digit on the right is greater than 5, drop it and increase the preceding digit by one, for the result 250.

243   Because the last digit on the right is less than 5, drop it and insert zero instead, for the result 240.

245   Because the last digit on the right is 5 or greater, drop it and increase the preceding digit by one.

As a general rule, the number of significant digits of the product or the division of two or more measurements should be no greater than that of the measurement with the least number of significant digits. For example, suppose the area of a table is to be determined. This is accomplished by measuring the length and width of the table, then multiplying one value by the other. In this procedure, it is inaccurate and meaningless to calculate and give an answer indicating greater accuracy than justified by the original data. For example: the length of a table is measured with a meter stick as 1.8245 m and the width as 0.3672 m. The area $A = 1.8245$ m $\times 0.3674$ m $= 0.6703213$ m^2 as shown on a calculator. This six-figure number is not justified as the correct area of the table as given by the two measurements. The correct value for the area is 0.6703 m^2. This value has four significant digits corresponding to the least number of significant digits in the original data.

When adding or subtracting significant figures use the following two rules:

Rule 1. A known digit plus or minus a doubtful digit will give a doubtful digit.
Rule 2. Only one doubtful digit is allowed in a significant figure.

For example:

2.34	the 4 is doubtful
+ 16.5	the 5 is doubtful
28.84	the 8 and the 4 are doubtful

From Rule 1, the known digit 3 plus the doubtful digit 5 equals a doubtful digit 8. Therefore, the correct answer is 28.8. This is a significant figure with only one doubtful number, which satisfies Rule 2.

Conversion of Units

The solution to many problems in physical science requires changing units of measurements from one system to another. The following method is simple and easy once it is understood. An understanding of the method is gained by solving a problem.

What is the distance in feet traveled by an automobile in 10 s, if the average speed of the car is 60 mi/h? Because we want to know the distance traveled in feet, we must convert miles per hour to feet per second. This can be done as follows:

1. Write down the term you want to convert and set it equal to itself. At the same time, make a mental or written note of the terms to which you wish to convert.

$$\frac{60 \text{ mi}}{\text{h}} = \frac{60 \text{ mi}}{\text{h}}$$

2. Multiply the right side of the equation by 1 or 1/1, which is the same thing.

$$\frac{60 \text{ mi}}{\text{h}} = \frac{60 \text{ mi}}{\text{h}} \times \frac{1}{1}$$

3. Place a unit in the numerator or denominator of the 1/1 that will cancel out an unwanted unit in the original term. In this case, we would place the unit "mile" in the denominator. At the same time, place an equal term above our newly introduced denominator unit, in order to keep the value equal to 1/1. Because we are after feet in our answer for this problem, use 5280 ft in the numerator of the 1/1, because that is equal to 1 mi. Then cancel like units.

$$\frac{60 \text{ mi}}{\text{h}} = \frac{60 \text{ \cancel{mi}}}{\text{h}} \times \frac{5280 \text{ ft}}{1 \text{ \cancel{mi}}}$$

4. The next step is to eliminate the unit hour in the denominator. This can be accomplished by multiplying again by 1/1 and placing the unit hour in the numerator of the 1/1 term. At the same time, an equal unit should be placed in the denominator of the 1/1 term in order to keep the value equal to 1/1. Because 60 min equal 1 h, we can use it and then cancel both "hour" terms.

$$\frac{60 \text{ mi}}{\text{h}} = \frac{60 \text{ \cancel{mi}}}{\cancel{\text{h}}} \times \frac{5280 \text{ ft}}{1 \text{ \cancel{mi}}} \times \frac{1 \cancel{\text{h}}}{60 \text{ min}}$$

5. At this point, we have feet per minute, but we want feet per second; therefore we must convert the minutes to seconds. Again we multiply by 1 or 1/1 and add the appropriate units for conversion. Because 60 s equal 1 min, we obtain

$$\frac{60 \text{ mi}}{\text{h}} = \frac{60 \text{ mi}}{\text{h}} \times \frac{5280 \text{ ft}}{1 \text{ mi}} \times \frac{1 \text{ h}}{60 \text{ min}} \times \frac{1 \text{ min}}{60 \text{ s}}$$

6. After all like units are canceled, the numbers are canceled insofar as is possible; we arrive at

$$\frac{60 \text{ mi}}{\text{h}} = \frac{528 \text{ ft}}{6 \text{ s}}$$

or, after dividing,

$$\frac{60 \text{ mi}}{\text{h}} = \frac{88 \text{ ft}}{\text{s}}$$

The problem can now be solved using 88 ft/s for 60 mi/h. Knowing that

$$d = vt$$

and substituting

$$d = \frac{88 \text{ ft}}{\text{s}} \times 10 \text{ s}$$

we obtain

$$d = 880 \text{ ft}$$

Experimental Error

The process of taking any measurement always involves some uncertainty. Such uncertainty is usually called experimental error. Two methods are used to calculate the amount of error: (1) When an accepted or standard value of the physical quantity is known, percent error is calculated. (2) When no accepted value exists, percent difference is calculated.

$$\text{Percentage error} = \frac{\text{absolute difference}^{*}}{\text{accepted value}} \times 100$$

$$= \frac{(E_v - A_v)}{A_v} \times 100$$

where E_v = experimental value and A_v = accepted value (known as standard value).

$$\text{Percentage difference} = \frac{\text{absolute difference}}{\text{average}} \times 100$$

where

$$\text{average} = \frac{E_1 + E_2 + E_3 + \ldots}{n}$$

E_1, E_2, and E_3 represent the experimental values and n represents the number of experimental values being averaged. Or this can be written as:

$$\text{Percentage difference} = \frac{(E_L - E_s)}{\text{average}} \times 100$$

where E_L = largest experimental value and E_s = smallest experimental value.

* Absolute difference means that the smaller value is subtracted from the larger value.

Conversion Factors

Mass
1 gram $= 10^{-3}$ kg $= 6.85 \times 10^{-5}$ slug
1 kg $= 10^3$ g $= 6.85 \times 10^{-2}$ slug
1 slug $= 1.46 \times 10^4$ g $= 14.6$ kg
1 amu $= 1.66 \times 10^{-24}$ g $= 1.66 \times 10^{-27}$ kg

Length
1 cm $= 10^{-2}$ m $= 0.394$ in
1 m $= 10^{-3}$ km $= 3.28$ ft $= 39.4$ in
1 km $= 10^3$ m $= 0.62$ mi
1 in $= 2.54$ cm $= 2.54 \times 10^{-2}$ m
1 ft $= 12$ in $= 30.48$ cm $= 0.3048$ m
1 mi $= 5280$ ft $= 1609$ m $= 1.609$ km
1A $= 10^{-10}$ m $= 10^{-8}$ cm
1 parsec $= 3.26$ light years $= 205{,}265$ a.u.

Time
1 h $= 60$ min $= 3600$ s
1 day $= 24$ h $= 1440$ min $= 8.64 \times 10^4$ s
1 year $= 365$ days $= 8.76 \times 10^3$ h
$\qquad = 5.26 \times 10^5$ min $= 3.16 \times 10^7$ s

Energy
1 joule $= 10^7$ erg $= 0.738$ ft $\cdot$ lb
$\qquad = 0.239$ cal $= 9.48 \times 10^{-4}$ Btu
$\qquad = 6.24 \times 10^{18}$ eV
1 kcal $= 4186$ J $= 4.186 \times 10^{10}$ erg
$\qquad = 3.968$ Btu
1 Btu $= 1055$ J $= 1.055 \times 10^{10}$ erg
$\qquad = 778$ ft $\cdot$ lb $= 0.252$ kcal
1 cal $= 4.186$ J $= 3.97 \times 10^{-3}$ Btu
$\qquad = 3.09$ ft $\cdot$ lb
1 ft $\cdot$ lb $= 1.36$ J $= 1.36 \times 10^7$ erg
$\qquad = 1.29 \times 10^{-3}$ Btu
1 eV $= 1.60 \times 10^{-19}$ J $= 1.60 \times 10^{-12}$ erg
1 kWh $= 3.60 \times 10^6$ J $= 3.413 \times 10^3$ Btu

Speed
1 m/s $= 3.6$ km/h $= 3.28$ ft/s $= 2.24$ mi/h
1 km/h $= 0.278$ m/s $= 0.621$ mi/h $= 0.911$ ft/s
1 ft/s $= 0.682$ mi/h $= 0.305$ m/s $= 1.10$ km/h
1 mi/h $= 1.467$ ft/s $= 1.609$ km/h $= 0.447$ m/s
60 mi/h $= 88$ ft/s

Force
1 newton $= 10^5$ dynes $= 0.225$ lb
1 dyne $= 10^{-5}$ N $= 2.25 \times 10^{-6}$ lb
1 lb $= 4.45 \times 10^5$ dynes $= 4.45$ N
Equivalent weight of 1 kg mass $= 2.2$ lb
$\qquad\qquad\qquad\qquad\qquad = 9.8$ N

Pressure
1 atm $= 14.7$ lb/in^2 $= 1.013 \times 10^5$ N/m^2
$\qquad = 1.013 \times 10^6$ dyne/cm^2
$\qquad = 30$ in Hg $= 76$ cm Hg
1 bar $= 10^6$ dyne/cm^2 $= 10^5$ Pa
1 millibar $= 10^3$ dyne/cm^2 $= 10^2$ Pa
1 Pa $= 1$ N/m^2 $= 10^{-2}$ millibar

Power
1 watt $= 0.738$ ft $\cdot$ lb/s $= 1.34 \times 10^{-3}$ hp
$\qquad = 3.41$ Btu/h
1 ft $\cdot$ lb/s $= 1.36$ W $= 1.82 \times 10^{-3}$ hp
1 hp $= 550$ ft $\cdot$ lb/s $= 745.7$ watt
$\qquad = 2545$ Btu/h